KNOT
PROJECTIONS

KNOT PROJECTIONS

Noboru Ito

The University of Tokyo
Meguro-ku, Tokyo, Japan

CRC Press
Taylor & Francis Group
Boca Raton London New York

CRC Press is an imprint of the
Taylor & Francis Group, an **informa** business
A CHAPMAN & HALL BOOK

CRC Press
Taylor & Francis Group
6000 Broken Sound Parkway NW, Suite 300
Boca Raton, FL 33487-2742

First issued in paperback 2020

© 2016 by Taylor & Francis Group, LLC
CRC Press is an imprint of Taylor & Francis Group, an Informa business

No claim to original U.S. Government works

ISBN-13: 978-1-4987-3675-6 (hbk)
ISBN-13: 978-0-367-65829-8 (pbk)

Visit the Taylor & Francis Web site at
http://www.taylorandfrancis.com

and the CRC Press Web site at
http://www.crcpress.com

To my teachers, family,
and friends.

Contents

CHAPTER 1 ▪ Knots, knot diagrams, and knot projections   1

  1.1   DEFINITION OF KNOTS FOR HIGH SCHOOL STUDENTS   1
  1.2   HOW TO DEFINE THE NOTION OF A KNOT DIAGRAM?   2
  1.3   KNOT PROJECTIONS   4
  1.4   TIPS: DEFINITION OF KNOTS FOR UNDERGRADUATE
        STUDENTS   4

CHAPTER 2 ▪ Mathematical background (1920s)   7

  2.1   REIDEMEISTER'S THEOREM FOR KNOT DIAGRAMS AND
        KNOT PROJECTIONS   8
  2.2   PROOF OF REIDEMEISTER'S THEOREM FOR KNOT
        DIAGRAMS   9
  2.3   PROOF OF REIDEMEISTER'S THEOREM FOR KNOT
        PROJECTIONS   11
  2.4   EXERCISES   13

CHAPTER 3 ▪ Topological invariant of knot projections (1930s)   17

  3.1   ROTATION NUMBER   17
  3.2   CLASSIFICATION THEOREM FOR KNOT PROJECTIONS
        UNDER THE EQUIVALENCE RELATION GENERATED BY
        $\overline{\Delta}$, RⅡ, AND RⅢ   19

CHAPTER 4 ▪ Classification of knot projections under RI and
             RⅡ (1990s)   29

  4.1   KHOVANOV'S CLASSIFICATION THEOREM   30
  4.2   PROOF OF THE CLASSIFICATION THEOREM UNDER RI
        AND RⅡ   31
  4.3   CLASSIFICATION THEOREM UNDER RI AND STRONG RⅡ
        OR RI AND WEAK RⅡ   37

4.4   CIRCLE NUMBERS FOR CLASSIFICATION UNDER RI AND STRONG RⅡ   39

4.5   EFFECTIVE APPLICATIONS OF THE CIRCLE NUMBER   44

4.6   FURTHER TOPICS   49

4.7   OPEN PROBLEMS AND EXERCISE   50

CHAPTER 5 ■ Classification by RI and strong or weak RⅡ (1996–2015)   53

5.1   AN EXAMPLE BY HAGGE AND YAZINSKI   53

5.2   VIRO'S STRONG AND WEAK RⅢ   55

5.3   WHICH KNOT PROJECTIONS TRIVIALIZE UNDER RI AND WEAK RⅢ?   57

5.4   WHICH KNOT PROJECTIONS TRIVIALIZE UNDER RI AND STRONG RⅢ?   61

5.5   OPEN PROBLEM AND EXERCISES   70

CHAPTER 6 ■ Techniques for counting sub-chord diagrams (2015–Future)   73

6.1   CHORD DIAGRAMS   73

6.2   INVARIANTS BY COUNTING SUB-CHORD DIAGRAMS   75

     6.2.1   Invariant X 75

     6.2.2   Invariant H 76

     6.2.3   Invariant λ 80

6.3   APPLICATIONS OF INVARIANTS   82

6.4   BASED ARROW DIAGRAMS   84

6.5   TRIVIALIZING NUMBER   86

6.6   EXERCISES   92

CHAPTER 7 ■ Hagge–Yazinski Theorem (Necessity of RI)   95

7.1   HAGGE–YAZINSKI THEOREM SHOWING THE NON-TRIVIALITY OF THE EQUIVALENCE CLASSES OF KNOT PROJECTIONS UNDER RI AND RⅢ   96

     7.1.1   Preliminary   96

     7.1.2   Box   97

     7.1.3   Structure of the induction   97

     7.1.4   Moves $1a$, $1b$, and RⅢ inside a rectangle   98

     7.1.5   Moves $1a$, $1b$, and RⅢ outside the rectangles   98

	7.1.5.1	1a	98
	7.1.5.2	1b	99
	7.1.5.3	R$\mathrm{I\!I\!I}$	99
7.2	ARNOLD INVARIANTS		100
7.3	EXERCISES		104

CHAPTER 8 ▪ Further result of strong (1, 3) homotopy — 107

8.1	STATEMENT	107
8.2	PROOF OF THE STATEMENT	108
8.3	OPEN PROBLEMS AND EXERCISE	117

CHAPTER 9 ▪ Half-twisted splice operations, reductivities, unavoidable sets, triple chords, and strong (1, 2) homotopy — 121

9.1	SPLICES	122
9.2	ITO-SHIMIZU'S THEOREM FOR HALF-TWISTED SPLICE OPERATIONS	124
9.3	UNAVOIDABLE SETS	125
9.4	UPPER BOUNDS OF REDUCTIVITIES	130
9.5	KNOT PROJECTION WITH REDUCTIVITY ONE: REVISITED	138
9.6	KNOT PROJECTION WITH REDUCTIVITY TWO	146
9.7	TIPS	150
9.7.1	Tip I	150
9.7.2	Tip $\mathrm{I\!I}$	150
9.8	OPEN PROBLEMS AND EXERCISES	151

CHAPTER 10 ▪ Weak (1, 2, 3) homotopy — 155

10.1	DEFINITIONS	156
10.2	STRONG (1, 2, 3) HOMOTOPY AND THE OTHER TRIPLES	156
10.3	DEFINITION OF THE FIRST INVARIANT UNDER WEAK (1, 2, 3) HOMOTOPY	158
10.4	PROPERTIES OF $W(P)$	161
10.5	TIPS	164
10.5.1	The trivializing number is even	164

10.5.2 One-half of the trivializing number is greater than or equal to the canonical genus of a knot projection   167

10.5.3 If $tr(P) = c(P) - 1$, then P is a $(2, p)$-torus knot projection   172

10.6 EXERCISE   174

CHAPTER 11 ▪ Viro's quantization of Arnold invariant   177

11.1 RECONSTRUCTION OF ARNOLD INVARIANT J^- 177

11.2 GENERALIZATION   185

11.3 QUANTIZATION   188

11.4 KNOT PROJECTION ON A SPHERE   190

11.5 EXERCISES   195

Index   199

Preface

If you see something, do you understand what it is?

You may be fortunate enough to possibly be able to grab it with your hand; however, simply holding something does not mean you understand it well. For example, what would you think about Figure 0.1. Figure 0.1 shows an example of a *knot diagram*.

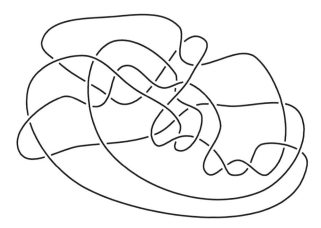

FIGURE 0.1 Knot diagram

In fact, there are many such things in the world that we can see and touch but may not understand what they are.* For some mathematicians, it is elementary to determine whether two representations corresponding to the same knot (= same closed string object), which are shown in Figure 0.2; Figure 0.2 shows two representations of the most simple non-trivial knot.

However, it would *not* be elementary to determine the same for the next pair as shown in Figure 0.3. In fact, this pair is called a version of the *Perko pair*† representing the same knot.

Historically, Perko found duplication in a knot table from the nineteenth century, which implies that no one pointed out the duplication for about 75

---

*The knot diagram in Figure 0.1 represents a trivial knot, which was constructed by referring to [1] and "The Culprit Undone" in [2, 4].

† "Perko pair" is a famous term among specialists. However, the author found several versions of the Perko pair and noticed that it may not represent a unique pair.

FIGURE 0.2   Two representations of the same knot

FIGURE 0.3   Perko pair

years. Hence, we can say that, in general, it is difficult to determine whether two knot diagrams truly represent two different knots or actually represent the same knot. Perko obtained a deformation (called an isotopy) from the left figure to the right figure, as shown in Figure 0.4.

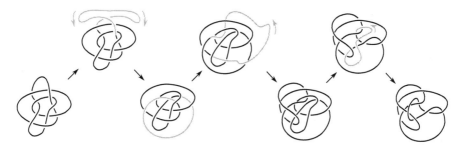

FIGURE 0.4 Isotopy from the left to the right figure

If we omit every over/under-information of a crossing of a knot diagram, the result is called a *knot projection*. Is it still difficult to determine what a knot projection is?

In the 1920s, Reidemeister obtained a result that showed that RI, RII, and RIII are sufficient to describe a deformation of any knot projection into a simple closed curve, where RI, RII, and RIII are local replacements between two knot projections as shown in Figure 0.5. Traditionally, in mathematics, such a deformation is treated as a notion of *homotopy*.

FIGURE 0.5 Reidemeister moves RI, RII, and RIII

In 2001, Östlund formulated a question as follows:
Östlund's question: Let RI and RIII denote the local replacements between two knot projections as shown in Figure 0.5. For every knot projection P, are RI and RIII sufficient to describe a deformation of P into a simple closed curve?

In 2014, Hagge and Yazinski were the first to obtain an example to answer to this question, which is shown in Figure 0.6.

However, a function on the set of the knot projections that can be used to easily detect which knot projection is related to a simple closed curve by a finite sequence generated by RI and RIII has not been found. We still do not have sufficient information about homotopies generated by RI and RIII for knot projections and more research on knot projections is needed.

The objective of this book to address the classification problem of knot projections and discuss related topics. This monograph consists of 11 chapters.

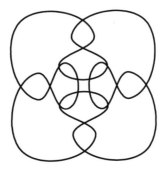

FIGURE 0.6 Example obtained by Hagge and Yazinski

Chapter 1 provides the definitions of knots and knot projections; in Chapter 2, we learn why we consider Reidemeister moves from a result obtained in the 1920s (Reidemeister's Theorem). In Chapter 3, we describe a useful result obtained by Whitney (1930s), i.e., two knot projections with the same rotation number on a plane if and only if they are related by a finite sequence generated by RⅡ, RⅢ, and a plane isotopy. In Chapter 4, we reach the end of the 20th century, where Khovanov's Theorem (1997) and its proof are described; this theorem can be used to obtain the classification of knot projections under homotopy generated by RI, RⅡ, and sphere isotopy.

After obtaining the basic results based on the studies of Reidemeister, Whitney, and Khovanov, we discuss most recent results, i.e., from 2013–2015. We consider classifications using refined Reidemeister moves from the latter part of Chapter 4 and Chapter 5. We decompose RⅡ into strong RⅡ and weak RⅡ; similarly, we decompose RⅢ into strong RⅢ and weak RⅢ. This decomposition is performed to not only provide a more detailed classification from a general one but also to open a new viewpoint of this mathematical area.

Historically, Arnold discussed classification theory using the singularity theory in mathematics to classify plane curves (= knot projections on a plane) under strong RⅡ and RⅢ, weak RⅡ and RⅢ, or RⅡ, i.e., without RI. The structure of a differential that is used to obtain a good classification of knot projection leads to the notion of *Legendrian knots*; however, RI cannot be used. By contrast, following Arnold's theory, many mathematicians have obtained excellent results, which implies a good understanding of a classification with respect to the equivalence relation generated by strong RⅡ and weak RⅢ. Many of these works provide ideas to study knot projections further.

Thus, finally, we study the classification theory of knot projections by considering a pair consisting of RI and other Reidemeister moves, denoted by (RI, ·), as shown in the following list. Each theory should be constructed depending on the situation of each case because there is difficulty when considering RI as mentioned above. In these chapters, we introduce ideas to obtain new invariants of knot projections. Here, an invariant of knot projections is a function on the set of the knot projections. Invariants are useful to determine which two knot projections are not equivalent. In other words, for a case where there is a situation such that there exist two different knot projections with the same invariant values, the classification problem is still open.

- (RI, strong RII) or (RI, weak RII): the latter part of Chapter 4,

- (RI, strong RIII): Chapter 5,

- (RI, strong RIII) or (RI, weak RIII): Chapter 6,

- (RI, RIII): Chapter 7,

- (RI, strong RIII) revisited: Chapter 8,

- (RI, weak RII, weak RIII): Chapter 10

Chapter 9 discusses topics related to (RI, strong RII), i.e., unavoidable sets. In topological studies, "unavoidable sets" is a famous term with respect to the four-color theorem. In Chapter 9, unavoidable sets for knot projections as described by Shimizu are introduced. Several definitions of reductivity related to unavoidable sets and characterizations are introduced, based on the works in [6] and [3].

In the last chapter, Chapter 11, we appreciate Viro's remarkable work, i.e., quantization of Arnold invariants of knot projections on a plane; his work obtains an infinite family of invariants under (weak RII, weak RIII). This is one of the most important goals achieved by Arnold's theory for plane curves. However, there has been no such quantum theory for knot projections with respect to (RI, ·) or a unified theory of invariants yet. Therefore, this chapter is a part of the epilogue of this book.

The author would like to express his appreciation to Yukari Funakoshi (Nara Women's University) for the beautiful artwork preparation based on the author's hand-drawn figures.

The author thanks the editor, Nair Sunil (CRC Press); editorial assistant, Alexander Edwards (CRC Press); his affiliate Waseda Institute for Advanced Study (WIAS) and its staff; and Senja Barthel (Imperial College London) for their valuable input.

Noboru Ito
Tokyo, February 2016

Bibliography

[1] M. Burton, http://mickburton.co.uk/2015/06/05/how-do-you-construct-haken-gordian-knot/.

[2] A. Henrich and L. H. Kauffman, Unknotting unknots, *Amer. Math. Monthly* **121** (2014), 379–390.

[3] N. Ito and Y. Takimura, Knot projections with reductivity two, *Topology Appl.* 193 (2015), 290–301.

[4] L. H. Kauffman and S. Lambropoulou, Hard unknots and collapsing tangles, *Introductory Lectures on knot theory*, 187–247, Ser. Knots Everything, 46, World Sci. Publ., Hackensack, NJ, 2012.

[5] B. Sanderson, Knot Theory Lectures, http://homepages.warwick.ac.uk/~maaac/MA3F2-page.html, Can you see the isotopy between the Perko pair?

[6] A. Shimizu, The reductivity of spherical curves, *Topology Appl.* **196** (2015), part B, 860–867.

Contributor

Yukari Funakoshi
(Artwork preparation)
Nara Women's University
Nara, Japan

CHAPTER 1

Knots, knot diagrams, and knot projections

CONTENTS

1.1   Definition of knots for high school students 1
1.2   How to define the notion of a knot diagram? 2
1.3   Knot projections 4
1.4   Tips: Definition of knots for undergraduate students 4

K̲not projections are knots in a three-dimensional space as closed string objects. Knots are not only common in the field of mathematics but also in some areas of theoretical physics and biology. In this chapter, we define knots, knot diagrams, and knot projections.

1.1 DEFINITION OF KNOTS FOR HIGH SCHOOL STUDENTS

A *knot* is the union of a finite number of straight line segments with no boundaries in $\mathbb{R}^3$ such that each endpoint joins exactly one of the remaining endpoints. Here, $\mathbb{R}^3$ is defined as $\{(x, y, z) \mid x, y, z \text{ are real numbers}\}$. We can use a smooth plane curve, called a *knot diagram*, to represent a knot (see Section 1.4), which is shown in Figure 1.1 (left). A straight line segment is called an *edge* and an endpoint is called a *vertex*.

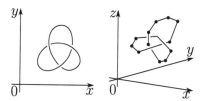

FIGURE 1.1 A knot (right) and its knot diagram (left)

Two knots are equivalent if there is a finite sequence generated by moving Δ between the two knots, where Δ denotes replacing one edge with two connected edges in three-dimensional space $\mathbb{R}^3$ as shown in Figure 1.2 or an inverse move such that there is no vertex or edge of the knot in the triangle corresponding to Δ.

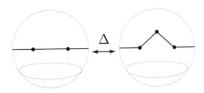

FIGURE 1.2 Δ: a local deformation in three-dimensional space

Here, note that by a finite sequence generated by Δ, we can add a vertex on an edge, as shown in Figure 1.3 ([1]).

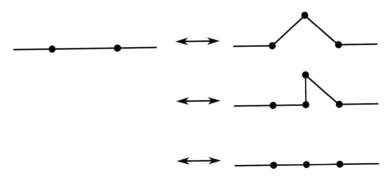

FIGURE 1.3   Inserting one vertex into an edge and removing the vertex

1.2   HOW TO DEFINE THE NOTION OF A KNOT DIAGRAM?

We have already defined a knot in the previous section. To study knots, we need to represent them. One way to represent them is in the form of a *knot diagram*. For a given knot, we consider its projection image onto a plane. When we consider such a projection, we should be careful so that the cases shown in Figure 1.4 do not occur.

Now, straight line segments on the plane are also called *edges* and two points on the boundary of each edge are called *vertices*. Let us consider the projection from $\mathbb{R}^3$ to a plane that satisfies the following rule:

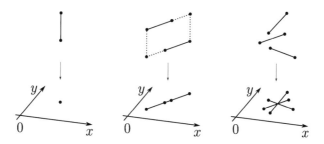

FIGURE 1.4 Forbidden cases

- Rule 1 : An edge should be projected onto an edge, not a vertex;

- Rule 2 : Every intersection should be a transverse double point;

- Rule 3 : Over-/underinformation of every double point is specified.

The projection image of a knot is called a *knot diagram on a plane*.

In our study, we consider knot diagrams not only on a plane but also knot diagrams on a sphere (cf. Chapters 2 and 3). Traditionally, we trace a knot diagram on an xy-plane onto a curve on the sphere, i.e., $\{(x, y, z) \mid x^2 + y^2 + (z-1)^2 = 1\}$ in the xyz-space $\mathbb{R}^3$. In this book, such a projection is called *stereographic projection* (Figure 1.5).

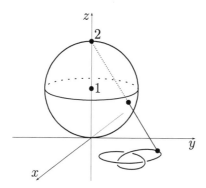

FIGURE 1.5 Stereographic projection

Consider a line connecting $(0, 0, 2)$ to a point on a knot diagram; then, proceed along the knot diagram and trace the intersection between the line and the sphere. This obtains the projection from a transverse double point on a plane to that of the sphere; using this, we can determine which path is near $(0, 0, 2)$ (Figure 1.6). The set of the intersection points on the sphere represents a *knot diagram on the sphere* and, in this book, we look at this set from an inner point of the sphere.

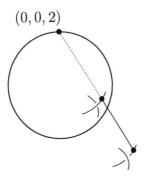

$(0, 0, 2)$

FIGURE 1.6 Over-/underinformation for a double point

1.3 KNOT PROJECTIONS

When we consider a natural projection (Figure 1.7) ignoring the over/under-information of every double point of a knot diagram (on a plane or on a sphere), the projection image is called a *knot projection* (on a plane or on a sphere, respectively). In other words, if we consider omitting Rule 3 when defining knot diagrams, we obtain a knot projection.

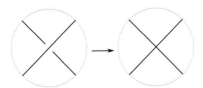

FIGURE 1.7 Projection from every crossing to the corresponding double point

1.4 TIPS: DEFINITION OF KNOTS FOR UNDERGRADUATE STUDENTS

A *knot* is an embedding of a circle into $\mathbb{R}^3$. Two knots K_0 and K_1 are *ambient isotopic* if there exists a continuous map $F(*, t)$ for $(x, t) \in \mathbb{R}^3 \times [0, 1]$ such that each $F(*, t) : \mathbb{R}^3 \to \mathbb{R}^3$, $F(K_0, 0) = K_0$, and $F(K_0, 1) = K_1$.

Knots defined in Section 1.1 are often called *piecewise linear knots*.

If a knot is ambient isotopic to a piecewise linear knot, then it is called a *tame knot*. If a knot is not a tame knot, it is called a *wild knot*. A standard wild knot theory has not been established yet. We hope that young researchers will contribute to building a wild knot theory. In this book, set the knots considered are only tame knots and thus, tame knots are simply called knots.

Bibliography

[1] L. H. Kauffman, Topics in combinatorial knot theory, http://homepages. math.uic.edu/ kauffman/KFI.pdf

CHAPTER 2

Mathematical background (1920s)

CONTENTS

2.1   Reidemeister's theorem for knot diagrams and knot projections   8
2.2   Proof of Reidemeister's theorem for knot   diagrams ...   9
2.3   Proof of Reidemeister's theorem for knot   projections ..   11
2.4   Exercises   13

A fundamental fact related to knot projections is that any two knot projections are related by a finite sequence generated by $\overline{\Delta}$, RI, RII, and RIII. The equivalence generated by $\overline{\Delta}$, i.e., the equivalence under a finite sequence generated by $\overline{\Delta}$, is called *sphere isotopy* (or *plane isotopy* when plane curves are considered). The equivalence relation generated by $\overline{\Delta}$, RII, and RIII is known as *regular homotopy* (i.e., RI is forbidden). The classification problem for the equivalence relation generated by $\overline{\Delta}$, RI, and RII (i.e., RIII is forbidden) was solved by Khovanov ([2], 1997). The case for which RII cannot be used is still unexplained; moreover, the knot projections that can be trivialized have not been determined yet!

A replacement of two edges with a single edge of a (piecewise linear) knot diagram or a knot projection is denoted by $\overline{\Delta}$. Its inverse is also denoted by $\overline{\Delta}$. Note that this local deformation occurs in a two-dimensional space; also note that the triangle may contain edges (possibly with vertices) that simply pass through the triangle, as shown in Figure 2.1.

FIGURE 2.1 $\overline{\Delta}$: Local deformation in a two-dimensional space (two cases)

2.1 REIDEMEISTER'S THEOREM FOR KNOT DIAGRAMS AND KNOT PROJECTIONS

Reidemeister's study ([4], 1927) claims the following theorem, Theorem 2.1, which is called *Reidemeister's Theorem*.

Theorem 2.1 *Suppose that an arbitrary knot has two knot diagrams D and D'. Then, there exists a finite sequence between D and D' generated by $\overline{\Delta}$, $\mathcal{RI}$, $\mathcal{RII}$, and $\mathcal{RIII}$ (Figure 2.2).*

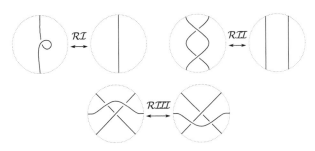

FIGURE 2.2 $\mathcal{RI}$, $\mathcal{RII}$, and $\mathcal{RIII}$

Using Reidemeister's theorem, we obtain the following theorem.

Theorem 2.2 *Two arbitrary knot projections are related by a finite sequence generated by $\overline{\Delta}$, RI, RII, and RIII (Figure 2.3).*

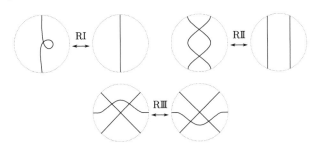

FIGURE 2.3 RI, RII, and RIII

In the rest of this book, a knot projection with no double points will be called a *trivial knot projection*.

Corollary 2.1 *Any knot projection can be related to a trivial knot projection by a finite sequence generated by $\overline{\Delta}$, $\mathcal{RI}$, $\mathcal{RII}$, and $\mathcal{RIII}$.*

We now consider that two knot projections P and P' are equivalent, which is denoted by $P = P'$, if they are related by a finite sequence generated by $\overline{\Delta}$.

According to Corollary 2.1, we can obtain any knot projection from a trivial knot projection using three local moves: RI, RII, and RIII. In other words, suppose that we have a trivial knot projection at time $t = 0$ and only three types of local moves are permitted. However, surprisingly, it is possible to obtain *any* kind of knot projection at $t = T$ for $t \leq T$.

What will happen if we restrict the moves that can be applied to a knot projection? Can we obtain knot projections in a similar manner? If we can determine a family of knots from just a knot projection, will it surprise us? Restricting the moves that can be applied to the knot projections opens new doors of interest.

For instance, it is natural to consider the following three cases:

- (Case 1) RI is forbidden.

- (Case 2) RII is forbidden.

- (Case 3) RIII is forbidden.

Case 1 is known as Whitney's theorem (1937) and Case 3 is known as Khovanov's theorem (1997). However, Case 2 is vague with a big open problem. The only thing we know is that the restriction that forbids the use of RII is non-trivial, i.e., there exists an RII decreasing the number of double points that cannot be generated by RI and RIII; this is shown in a theorem obtained by Hagge and Yazinski in 2014. Chapter 3 describes Whitney's theorem and Chapter 4 describes Khovanov's theorem. Chapter 7 describes Hagge–Yazinski's theorem. Chapters 5 and 6 describe the works of Ito (the author), Takimura, and Taniyama related to the problems in Case 2.

In this chapter, we describe the proof of Theorem 2.1 and Theorem 2.2 through the approaches followed in [3] and [1].

2.2 PROOF OF REIDEMEISTER'S THEOREM FOR KNOT DIAGRAMS

It is sufficient to show that a single Δ can be represented by a finite sequence generated by $\overline{\Delta}$, $\mathcal{RI}$, $\mathcal{RII}$, and $\mathcal{RIII}$.

We set the symbols for the preliminary of the proof. A single Δ indicates replacing an edge e_1 with two connected edges $e_2 \cup e_3$. Recall that the inverse replacement is also denoted by Δ. Then, two points, $e_1 \cap (e_2 \cup e_3)$, can be denoted by A and B as shown in Figure 2.4. When there is no ambiguity, we use A and B for the corresponding triangle on a knot diagram.

For describing several patterns in a simple manner, we do not include the over/under-information for every double point in this proof. This is because we can easily recover every possible case if we obtain the over/under-information for every double point. By definition of Δ, a triangle (its boundary and its

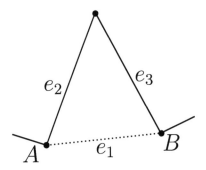

FIGURE 2.4 *A* and *B*

interior) concerned with Δ cannot penetrate any other part of a knot. Thus, the boundary of the triangle concerned with (projected) Δ has no edge that is tangential with respect to any other edge or vertex (except for an edge connecting to A or B) of the knot diagram (condition ($\ast$)). Without much effort, we can remove the extra vertices on the edges of the knot (recall Figure 1.3).

First, we divide the image of the projection of a single Δ into a finite number of triangles (Figure 2.5).

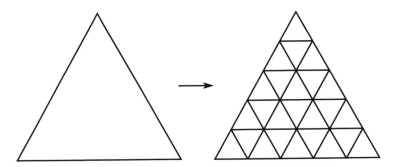

FIGURE 2.5 Decomposition

Second, we select a triangle and divide it into a finite number of triangles. From this process, we obtain the decomposition of a single Δ into a finite number of Δ until the following condition is satisfied: every triangle has either at most one vertex or at most one double point in its interior ($\ast$). We denote by Δ' a single Δ while condition ($\ast$) is satisfied after performing the decomposition. It is possible to obtain a decomposition of a given single Δ because we can avoid a finite number of forbidden cases. A single Δ consists of a finite number of local moves, each of which is labeled as Δ'. Thus, in the following, we can suppose that every Δ satisfies conditions ($\ast$) and ($\star$).

Under this condition (∗) and (⋆), when we look at a single Δ', we have exactly two cases, Case (1) and Case (2).

Case (1): If there exists an edge from A or B to the interior of the triangle, then for this Δ, we can construct a sequence generated by $\overline{\Delta}$ and $\mathcal{RI}$, or $\overline{\Delta}$, $\mathcal{RI}$, and $\mathcal{RIII}$ (Figure 2.6).

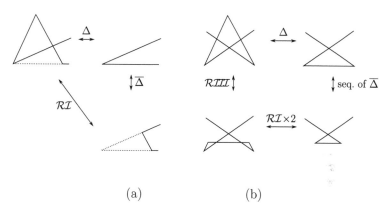

(a)   (b)

FIGURE 2.6   Case (1)

Case (2): If there is no edge from A or B to the interior of the triangle, we have the cases shown in Figure 2.7. In each of these cases, we can construct a finite sequence generated by $\overline{\Delta}$, $\mathcal{RII}$, and $\mathcal{RIII}$.

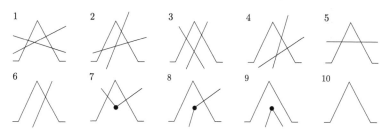

FIGURE 2.7   Case (2)

2.3 PROOF OF REIDEMEISTER'S THEOREM FOR KNOT PROJECTIONS

See Figure 2.8. For a given knot projection, we always construct a knot diagram of a trivial knot (see Figure 2.8). First, we select a point, called a *base point*, on the knot projection, which is not a double point; we then choose an orientation of the knot projection. Second, starting from the base point,

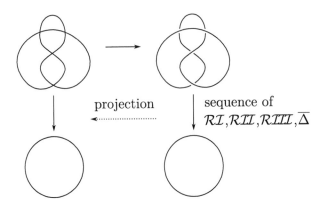

FIGURE 2.8 Construction of a trivial knot from a knot projection (descending construction).

we proceed the knot projection according to its orientation until we return to the base point. Consider encountering a double point that consists of paths b_1 and b_2. The path on which we encounter a double point first (resp. second) is denoted by b_1 (resp. b_2). Third, we set path b_1 (resp. b_2) as the over- (resp. under-) path for the double point. For every double point, we set the over-/under-information. As a result, we obtain a knot diagram of a trivial knot because this knot is a string object descending from the base point to path b_2 of the last double point (Figure 2.9). The knot diagram of this construction is denoted by D, which represents a trivial knot. Then, there exists

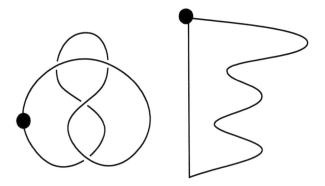

FIGURE 2.9 Construction of a knot diagram of a trivial knot

a finite sequence between D and the knot diagram with no double points such that the finite sequence is generated by $\overline{\Delta}$, $\mathcal{RI}$, $\mathcal{RII}$, and $\mathcal{RIII}$ by using Theorem 2.1. Therefore, neglecting over-/underinformation, we can obtain a finite sequence generated by $\overline{\Delta}$, RI, RII, and RIII.

2.4   EXERCISES

1. See the proof of Theorem 2.1 and consider Case (1), which is shown in Figure 2.6 (b). Consider whether this case can be decomposed into a sequence of other cases by appropriately triangulating Δ further.

2. Consider whether the proof of Theorem 2.1 can be rewritten by omitting Cases 7–9 in Figure 2.7 from the proof.

3. Consider whether Theorem 2.1 can be proved without using piecewise linear knot diagrams and instead using smooth knot diagrams; that is, smooth immersions into S^2 with data corresponding to the double points (over-/under- information of the double points).

4. Find a knot projection that cannot be trivialized by any finite sequence generated by $\overline{\Delta}$, RI, and RIII.

Bibliography

[1] L. H. Kauffman, Topics in combinatorial knot theory, `http://homepages.` `math.uic.edu/ kauffman/KFI.pdf`.

[2] M. Khovanov, Doodle groups, *Trans. Amer. Math. Soc.* **349** (1997), 2297–2315.

[3] V. V. Prasolov and A. B. Sossinsky, Knots, links, braids and 3-manifolds, An introduction to the new invariants in low-dimensional topology. Translations of Mathematical Monographs, **154**. *American Mathematical Society, Providence, RI,* 1997. viii+239 pp.

[4] K. Reidemeister, Elementare Begruendung der Knotentheorie, *Abh. Math. Sem. Univ. Hamburg* **5** (1927), 24–32.

CHAPTER 3

Topological invariant of knot projections (1930s)

CONTENTS

3.1   Rotation number 17
3.2   Classification theorem for knot projections under the
      equivalence relation generated by $\overline{\Delta}$, RⅡ, and RⅢ 19

R otation number is a complete invariant for the classification of plane curves that are equivalent to knot projections on a plane. The rotation number was introduced by Whitney in 1937 [2]. The same notion corresponding to the rotation number may have been introduced by other authors; however, we quote the fact from Arnold's book [1] here. Although we do not simply follow the contents of Whitney's paper, we provide the proof of the classification theorem with respect to the context in this book.

3.1   ROTATION NUMBER

Recall a knot diagram on a sphere. When we do not specify the over-/underinformation of two paths of a double point, we call it a knot projection.

Next, let us consider a knot diagram on a plane. When we do not specify the over-/underinformation of two paths of a double point, we call it a *plane curve*. In the situation that we consider a knot diagram of a piecewise linear knot, if necessary, we specify a piecewise linear plane curve.

A *rotation number* for a plane curve is described as follows. Let us select a starting point that is not a vertex. Next, we consider going along the curve starting at this starting point. In the following, let us determine how to change the angle from a point on the curve while going along the curve. Note that when we return to the starting point, the direction should be the same as the starting direction. Hence, the total difference between the starting angle and the ending angle is $k\pi$ $(k \in \mathbb{Z})$. Note that, for a piecewise linear plane curve, it is sufficient to calculate the difference between the angles at every vertex, i.e., for every vertex v, we can write the difference between the angles as ang(v) ($0 \leq$ ang(v) $< \pi$) and the sum $\frac{1}{2\pi} \sum_v$ ang(v) corresponds to the rotation number. The result $|k|$ is the rotation number of the plane curve.

Remark 3.1 It is well known that the rotation number is also defined for the smooth plane curves. The details are left to the readers.

More formally, we define the rotation number as follows. A given plane curve consists of a finite sequence of edges

$$e_1, e_2, e_3, \ldots, e_l$$

where two adjacent edges e_i and e_{i+1} have a common endpoint, i.e., vertex $v_{i,i+1}$. We assign a unit vector u_i to each e_i such that it has the same direction as that of e_i. For each $v_{i,i+1}$, the difference between u_i and u_{i+1} angles is ang($v_{i,i+1}$) ($-\pi$ ang π). Let C be a plane curve; then we define rot(C) by

$$\text{rot}(C) = \left| \frac{1}{\pi} \sum_{v_{i,i+1}} \text{ang}(v_{i,i+1}) \right|.$$

We will now familiarize ourselves a bit more with the rotation number. Because the rotation number is an integer, we do not need to consider all the values of ang. In other words, it is sufficient to check when the unit vector moves toward the north. Thus, we can select $v_{i,i+1}$ when moving toward the north. The north is usually defined by $(0, 1)$ when a plane is defined by an xy-plane $\mathbb{R}^2$. For every selected value of vertex $v_{i,i+1}$, denoted by $\tilde{v}_{i,i+1}$, we assign the sign $\epsilon(\tilde{v}_{i,i+1})$ as follows:

$$\epsilon(\tilde{v}_{i,i+1}) = \begin{cases} -1 \text{ if the unit vector moves toward the north from left to right} \\ 1 \text{ if the unit vector moves toward the north from right to left} \end{cases}$$

Thus, we obtain the formula

$$\text{rot}(C) = \left| \sum_{\tilde{v}} \epsilon(\tilde{v}_{i,i+1}) \right|. \tag{3.1}$$

In Chapter 2, Reidemeister moves RI, RII, and RIII for knot projections on a sphere are defined. However, Reidemeister moves are replacements in the local disk; thus, we can consider RI, RII, and RIII to be local moves on a plane. Here, from the above formula (3.1), we easily notice the following.

Theorem 3.1 *RII and RIII applied on a plane do not change the rotation number.*

Proof 3.1 Two disks D_1 and D_2 define RII. Four points (two pairs of points) on the boundary ∂D_1 of D_1 do not change under RII. Two points, say p_1 and p_2, of a pair have the same unit vector. Thus, $\sum \epsilon(\tilde{v}_{i,i+1})$ in the disk of RII is 0. If not, a simple arc connecting p_1 and P_2 has at least one double point, which is a contradiction. The same discussion for RⅢ implies the invariance of the rotation number under RⅢ. Six points (three pairs) on the boundary are fixed under a single RⅢ. Two points, say p_1 and p_2, of a pair have the same unit vector. A simple arc connecting p_1 and p_2 has no double point, which implies the claim.

3.2 CLASSIFICATION THEOREM FOR KNOT PROJECTIONS UNDER THE EQUIVALENCE RELATION GENERATED BY $\overline{\Delta}$, RⅡ, AND RⅢ

We can generalize Theorem 3.1 to Theorem 3.2.

Theorem 3.2 *C and C' can be related by a finite sequence generated by $\overline{\Delta}$, RⅡ, and RⅢ if and only if C has the same rotation number as that of C'.*

In other words, every plane curve can be related to exactly one element C_k, as shown in Figure 3.1.

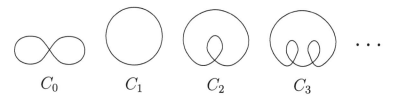

C_0 C_1 C_2 C_3 $\cdots$

FIGURE 3.1 C_k

Before starting with the proof, we define and recall some symbols.

Definition 3.1 Let C and C' be plane curves. We write $C \overset{\mathrm{RII,RIII}}{\sim} C'$ if C and C' are related by a finite sequence generated by $\overline{\Delta}$, RⅡ, and RⅢ.

Recall that we write $P = P'$ if two plane curves (i.e., knot projections on a plane) P and P' are related by a finite sequence generated by $\overline{\Delta}$.

Proof 3.2 First, we show that claim $(\star)$ implies the following:

$$C \overset{\mathrm{RII,RIII}}{\sim} C' \Leftrightarrow \mathrm{rot}(C) = \mathrm{rot}(C').$$

Claim $(\star)$: For a given plane curve C, there exists $C_* \in \{C_k\}$ such that $C \overset{\mathrm{RII,RIII}}{\sim} C_*$.

$((\Rightarrow)$ of $C \overset{\mathrm{RII, RIII}}{\sim} C' \Leftrightarrow \mathrm{rot}(C) = \mathrm{rot}(C').)$ Suppose that $C \overset{\mathrm{RII,RIII}}{\sim} C'$. By Theorem 3.1, $\mathrm{rot}(C) = \mathrm{rot}(C')$.

$((\Leftarrow)$ of $C \overset{\mathrm{RII, RIII}}{\sim} C' \Leftrightarrow \mathrm{rot}(C) = \mathrm{rot}(C').)$ Suppose that $\mathrm{rot}(C) = k$. By

Theorem 3.1 and $(\star)$, $k = \mathrm{rot}(C) = \mathrm{rot}(C_*)$; then, $C_* = C_k$. Thus, $C \overset{\mathrm{RII,RIII}}{\sim}$ C_k. Similarly, $\mathrm{rot}(C') = k$, $C' \overset{\mathrm{RII,RIII}}{\sim} C_k$. Hence, $C \overset{\mathrm{RII,RIII}}{\sim} C'$.

(End of proof of $(\star)$ implies $C \overset{\mathrm{RII,\ RIII}}{\sim} C' \Leftrightarrow \mathrm{rot}(C) = \mathrm{rot}(C')$)

Second, we show that the claim $(\star)$ implies the latter part of Theorem 3.2. Suppose $(\star)$, then $C \overset{\mathrm{RII,RIII}}{\sim} C_* \in \{C_k\}$. Then, we have

$$C \overset{\mathrm{RII,RIII}}{\sim} C' \Leftrightarrow \mathrm{rot}(C) = \mathrm{rot}(C').$$

Set $C' = C_*$ for C_* that appears in $(\star)$. If $\mathrm{rot}(C) = k$, $C_* = C_k$. This is because if $C_* = C_m$ $(m \neq k)$, $k = \mathrm{rot}(C) = \mathrm{rot}(C_*) = \mathrm{rot}(C_m) = m$, which is a contradiction. This implies the latter part of Theorem 3.2.

Next, we show claim $(\star)$. For this, we introduce the notion of a *teardrop disk*.

Definition 3.2 For a plane curve, consider starting at double point d and going along the curve until we return to the same point d. Suppose that the sub-curve obtained by starting and ending at d can be represented as disk D with the boundary, as shown in Figure 3.2. Then, D is called a *teardrop disk* and double point d is called the *teardrop-origin*.

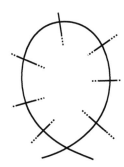

FIGURE 3.2 Teardrop disk

Lemma 3.1 *For a plane curve, there exists at least one teardrop disk.*

Proof 3.3 First, we note that when going along a plane curve starting at a double point p, we must return to the same double point p. This sub-curve tracked by starting and ending at p is denoted by $t(p)$. If double point p_0 is not the teardrop-origin, sub-curve $t(p_0)$ will have another double point p_1 (e.g., see Figure 3.3).

Next, we consider $t(p_1)$. If p_1 is not the teardrop-origin, the sub-curve $t(p_1)$ will have another double point. Then, the discussion can be repeated. We have a sequence:

$$p_0, p_1, p_2, \cdots$$

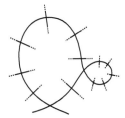

FIGURE 3.3 p_1 on $t(p_0)$

and

$$t(p_0), t(p_1), t(p_2), \ldots.$$

Note that by definition, any two sub-curves $t(p_i)$ and $t(p_j)$ may have common double points but cannot have common one-dimensional sub-curves (condition (∗)).

   Now we consider a graph with vertex p_i, where there exists an edge between p_i and p_j if $t(p_i)$ has vertex p_j (e.g., see Figure 3.4).

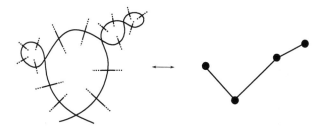

FIGURE 3.4 Sequence consisting of $\{t(p_i)\}$ and the corresponding graph

By condition (∗), this graph is a tree (i.e., it is connected and has no cyclic paths). Thus, this graph has an endpoint that corresponds to some p_k. Then $t(p_k)$ is a teardrop disk.

   In the following, by using the notion of *teardrop disks*, we show claim (⋆). For a given C, there exists a teardrop disk with a teardrop-origin p (Lemma 3.1). Consider a pulling operation (see Figure 3.5) performed on a teardrop disk at p until a sub-curve $\widetilde{t(p)}$ that contains only the teardrop-origin p (i.e., it corresponds to a simple circle with a boundary that has no sub-curves within its interior) is obtained; this sub-curve is called a *simple teardrop disk*. This process can be written by a finite sequence S generated by Δ, RI, RII, and RIII. However, sequence S has no RI because $t(p)$ is a teardrop disk.

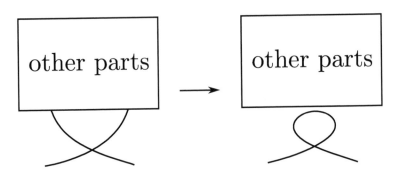

FIGURE 3.5 Pulling operation

If $t(p)$ has more than one double point, the process will eliminate at least one double point. Suppose that every teardrop disk is modified to obtain a

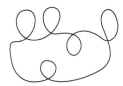

FIGURE 3.6 Curve consisting of simple teardrop disks

simple teardrop disk (e.g., see Figure 3.6). Then, using local moves in Figure 3.7, we have the claim ($\star$).

Next, we consider the corresponding result for knot projections on a sphere. When we make one hole at a point on the sphere, we can consider the punctured sphere to be a plane (or a disk). What does this hole represent? Traditionally, the hole is interpreted as an infinite point (cf. Chapter 1). Thus, if we can detect the local move passing through an infinite point, then we write the following statement. Recall that a knot projection on a plane is called a *plane curve*.

Let P be a knot projection on a sphere. Let C_P be a plane curve defined by selecting a point, denoted by ∞, in region $R(\infty)$ that is from $S^2 \setminus P$. $R(\infty)$ can be considered to be a disk surrounding C_P. Now we denote an arc that lies on the boundary of $R(\infty)$ as α. Thus, for C_P and $R(\infty)$, we define a replacement Ω_α of α in $R(\infty)$, as shown in Figure 3.8. If arc α is not specified, symbol "α" is omitted and the replacement can be written simply as Ω.

Proposition 3.1 *Let P and P' be knot projections on a sphere. $P \overset{RII,\ RIII}{\sim} P'$ on a sphere if and only if C_P and $C_{P'}$ are related by a finite sequence generated by $\overline{\Delta}$, RII, $RIII$, and Ω on a plane.*

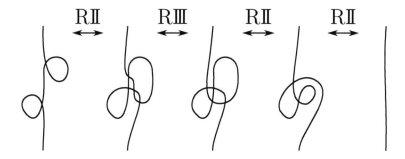

FIGURE 3.7 Sequence generated by RⅡ and RⅢ

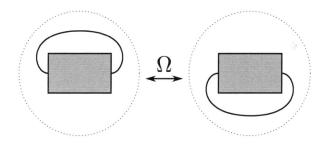

FIGURE 3.8 Replacement Ω

Proof 3.4 Suppose that $P \overset{\text{RⅡ, RⅢ}}{\sim} P'$. Thus, there is a finite sequence generated by RⅡ, RⅢ, $\overline{\Delta}_x$, and $\overline{\Delta}_\infty$, where if $\overline{\Delta}$; passes through a selected point, ∞, on a sphere, then $\overline{\Delta}$ is denoted by $\overline{\Delta}_\infty$, otherwise, it is denoted by $\overline{\Delta}_x$. Besides $\overline{\Delta}_\infty$, all other operations, i.e., RⅡ, RⅢ, and $\overline{\Delta}_x$, are not related to point ∞. Hence, we regard them as local moves RⅡ, RⅢ, and $\overline{\Delta}$ on a plane. Conversely, RⅡ, RⅢ, and $\overline{\Delta}$ on a plane are regarded as RⅡ, RⅢ, and $\overline{\Delta}_x$ on a sphere. Thus, every $\overline{\Delta}_\infty$ corresponds to a Ω and every Ω corresponds to $\overline{\Delta}_\infty$ (Figure 3.9), which implies the claim.

Let C_P and C' be plane curves such that C_P is obtained from C' by applying a single Ω to C'. In the following, we compute the difference between the rotation numbers $\text{rot}(C_P)$ and $\text{rot}(C')$. We give an arbitrary orientation to C_P and consider an operation on a double point of a plane curve C_P, called a *Seifert splice*. When we apply a Seifert splice to every double point, we obtain an arrangement of finite number of circles with no double points. We call this arrangement the *Seifert circle arrangement*, denoted by $S(C)$ for a plane curve C. The number of circles in $S(C)$ is denoted by $|S(C)|$. Let J be a simple closed curve in the Seifert circle arrangement. Let $\epsilon(J) = 1$

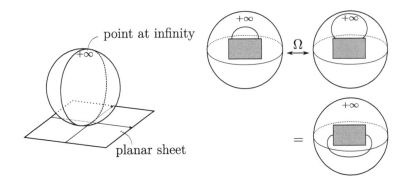

point at infinity

planar sheet

Ω

$=$

FIGURE 3.9 A point at infinity and Ω

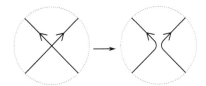

FIGURE 3.10 Seifert splice

(resp. $\epsilon(J) = -1$) if J is oriented counterclockwise (resp. clockwise). As in the example in Figure 3.11, we reformulate $\text{rot}(C_P)$ as

$$\text{rot}(C_P) = \left| \sum_{J \text{ in } S(C_P)} \epsilon(J) \right|.$$

A single Ω changes $\text{rot}(C_P)$ by exactly ± 2 and does not affect $|S(C_P)|$. We notice that $\text{rot}(C_P)$ is even (resp. odd) if and only if $|S(C_P)|$ is even (resp. odd).

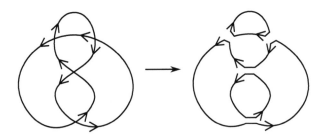

FIGURE 3.11 Seifert circle arrangement

Now we define the rotation number rot_S for knot projection P on a sphere as follows. Let C_P be a plane curve obtained by selecting an arbitrary point from $S^2 \setminus P$. We define $\text{rot}_S(P)$ by

$$\text{rot}_S(P) = \begin{cases} 0 & S(C_P) : \text{even} \\ 1 & S(C_P) : \text{odd}. \end{cases} \tag{3.2}$$

Theorem 3.3 *Let P and P' be knot projections on a sphere.*

$$P \overset{RII,\ RIII}{\sim} P' \Leftrightarrow \text{rot}_S(P) = \text{rot}_S(P').$$

Proof 3.5 ($\Rightarrow$) By Proposition 3.1, we have the claim using the definition of rot_S.

($\Leftarrow$) Conversely, suppose that $\text{rot}_S(P) = \text{rot}_S(P')$. Let P_0 (resp. P_1) be a knot projection such that C_P is C_0 (resp. C_1). Let P be an arbitrary knot projection on a sphere. It is sufficient to show that every plane curve C_P is related to either C_0 or C_1 by a finite sequence generated by $\overline{\Delta}$, RII, RIII, and Ω (**), which implies that $P \overset{RII,\ RIII}{\sim} P_i$ $(i = 0,\ 1)$.

If we have (**) as shown above,

$$\text{rot}_S(P) = \text{rot}_S(P') \Rightarrow P \overset{RII,\ RIII}{\sim} P_i \overset{RII,\ RIII}{\sim} P'.$$

This is because if P (resp. P') $\overset{RII,\ RIII}{\sim} P_i$ (resp. P_j) and $i \neq j$, $\text{rot}_S(P) = \text{rot}_S(P_i) \neq \text{rot}_S(P_j) = \text{rot}_S(P')$.

Now we show (**). Recall that there is a one-to-one correspondence between Ω and $\overline{\Delta}_\infty$. Every plane curve C_P is related to either C_0 or C_1 by a finite sequence generated by $\overline{\Delta}$, RII, RIII, and Ω. By the proof of Theorem 3.2, we have (*). Thus, it is sufficient to show that C_k is related to either C_0 or C_1 by a finite sequence generated by $\overline{\Delta}$, RII, RIII, and Ω. However, there can also be a sequence generated by $\overline{\Delta}$, RII, RIII, and Ω between C_k and C_{k-2} that can be easily seen from the case when $k = 4$, which can then be generalized for any value of k (Figure 3.12). Therefore, we have

$$C_k \to C_{k-2} \to \cdots \to C_3 \to C_1$$

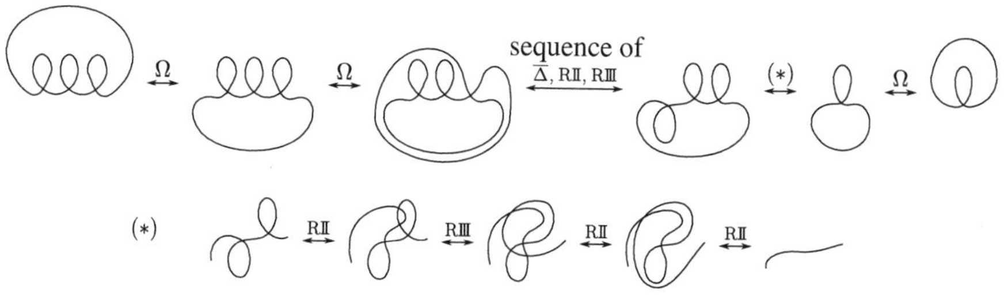

FIGURE 3.12 C_k to C_{k-2} (for $k = 4$)

or

$$C_k \to C_{k-2} \to \cdots \to C_4 \to C_2.$$

As in the rightmost move in the first line in Figure 3.12, a single Ω is used as a move between C_2 and C_0. Thus, we have the claim.

Bibliography

[1] V. I. Arnold, "Topological invariants of plane curves and caustics." Dean Jacqueline B. Lewis Memorial Lectures presented at Rutgers University, New Brunswick, New Jersey. University Lecture Series, 5. *American Mathematical Society, Providence, RI,* 1994. viii+60 pp.

[2] H. Whitney, On regular closed curves in the plane, *Compositio Math.* 4 (1937), 276–284.

CHAPTER 4

Classification of knot projections under RI and RII (1990s)

CONTENTS

4.1   Khovanov's classification theorem 30
4.2   Proof of the classification theorem under RI and RII 31
4.3   Classification theorem under RI and strong RII or RI and
        weak RII 37
4.4   Circle numbers for classification under RI and strong RII   39
4.5   Effective applications of the circle number 44
4.6   Further topics 49
4.7   Open problems and Exercise 50

Classification of knot projections under RI and RII was obtained in the latter half of the 20th century ([8], 1997). Originally, Khovanov obtained the classification by considering doodles. The notion of doodles was introduced by Fenn and Taylor ([2], 1979). Extending the notion of doodles, Khovanov showed the existence and uniqueness of the generators of knot projections on a plane. This chapter discuss Khovanov's classification theorem for knot projections on a sphere in the manner described in Ito–Takimura ([4], 2013). In this chapter, knot projections imply knot projections on a sphere unless specified otherwise. Recently, a classification theorem under RI and strong RII (resp. weak RII) was obtained by Ito–Takimura [5] by extending their theorem developed in 2013 [4]. Ito–Takimura [5] also obtained a new additive integer-valued invariant $|\tau(P)|$ of a knot projection P, called *circle number*, under RI and strong RII. The other known additive integer-valued invariant $a(P)$ of a knot projection under RI and strong RII was obtained by Arnold ([1], 1994), but $|\tau(P)|$ and $a(P)$ are independent of each other. The other additive integer-valued invariants under RI and strong RII are unknown.

Notation 4.1 When we go along a knot projection, the path starting from a double point to the next double point is called an *arc*. By definition, the boundary of an arc consists of one or two double point(s). Consider a disk, the boundary of which consists of n-arcs of a knot projection. The boundary of the disk is called an *n-gon*.

In the rest of this book, we will freely use the term "n-gon."

4.1 KHOVANOV'S CLASSIFICATION THEOREM

Theorem 4.1 ([8], [4]) *Suppose that two knot projections P and P' are related by a finite sequence generated by $\overline{\Delta}$, RI, and RII. Let 1a (resp. 2a) be an RI (resp. RII) that increases the number of double points. There exists a unique knot projection P^r with minimal number of double points such that P (resp. P') is obtained from P^r by a finite sequence generated by $\overline{\Delta}$, 1a, and 2a.*

Corollary 4.1 *P^r is a knot projection with no 1-gons and 2-gons.*

Using the symbols, we can state the above facts in a simple manner as Theorem 4.2.

Definition 4.1 For two knot projections P and P', we write $P \overset{RI, \, RII}{\sim} P'$ if there is a finite sequence generated by $\overline{\Delta}$, RI, and RII between P and P'.

Definition 4.2 For a knot projection P, denote P^r by a knot projection with no 1-gons and 2-gons obtained by a finite sequence generated by 1b and 2b, where 1b (resp. 2b) denotes RI (resp. RII) decreasing the number of double points. As stated in Theorem 4.1, RI (resp. RII) increasing the number of double points is denoted by 1a (resp. 2a) (see Figure 4.1).

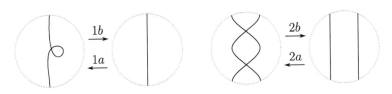

FIGURE 4.1 RI and RII

Next, for simplifying our discussion, if two knot projections P and P' are related by a finite sequence generated by $\overline{\Delta}$ (Figure 4.2), we say that P and P' are sphere isotopic and write $P = P'$. Instead of Theorem 4.1, we can state Theorem 4.2, simply.

Theorem 4.2 ([8], [4])

$$P \overset{RI, \, RII}{\sim} P' \Leftrightarrow P^r = P'^r.$$

FIGURE 4.2 $\overline{\Delta}$

4.2 PROOF OF THE CLASSIFICATION THEOREM UNDER RI AND RII

Now we start the proof of Theorem 4.2. First, ($\Leftarrow$) is simple to prove. If $P^r = P''^r$, by definition, $P \overset{\text{RI, RII}}{\sim} P^r = P''^r \overset{\text{RI, RII}}{\sim} P'$. Thus, we have the claim.

Next, we show ($\Rightarrow$). For the proof, we first prove Lemma 4.1 as below.

Lemma 4.1 ([8], [4]) *Let P_0 and P be knot projections such that $P_0 = P_0{}^r$ and $P_0 \overset{\text{RI, RII}}{\sim} P$. Then, there exists a finite sequence generated by $1a$ and $2a$ from P_0 to P.*

If we have Lemma 4.1, we can prove ($\Rightarrow$) of Theorem 4.2 as follows. If $P \overset{\text{RI, RII}}{\sim} P'$, by Lemma 4.1, there is a finite sequence generated by $1a$ and $2a$ from P''^r to P^r. Now assume that $P^r \neq P''^r$ (∗). Then, P^r has at least one 1-gon or 2-gon. However, by definition, P^r has no 1-gon or no 2-gon, which is a contradiction. Thus, the assumption (∗) is false and $P''^r = P^r$.

In the following, we show Lemma 4.1. By assumption, there exists a finite sequence generated by RI and RII from P_0 to P (up to a finite number of $\overline{\Delta}$). We focus on the first occurrence of $1b$ or $2b$, called a *b-move*, in the sequence (◇) from P_0 to P. We can consider at most four cases for (◇) as follows.

- Case 1: $\cdots\cdots (1a)(1b)$

- Case 2: $\cdots\cdots (2a)(1b)$

- Case 3: $\cdots\cdots (1a)(2b)$

- Case 4: $\cdots\cdots (2a)(2b)$

For each of these four cases, we show that if the b-move is the nth move, it can be moved to the $(n-1)$th move or it can be eliminated.
• Case 1. We can draw two successive local figures corresponding to $(1a)(1b)$ as shown in Figure 4.3. Denote by x (resp. y) a disk with a boundary generated (resp. eliminated) by $1a$ (resp. $1b$).
• Case 1-(i) where $\partial x \cap \partial y \neq \emptyset$. In this case, there can be two situations, as shown in Figure 4.4. For each situation, two successive local moves $(1a)(1b)$ can be replaced with $\emptyset$.
• Case 1-(ii) where $\partial x \cap \partial y = \emptyset$. In this case, for two successive local moves $(1a)(1b)$, depending on the pair, we can have a bypass $(1b)(1a)$ instead of the

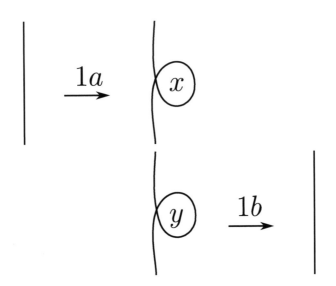

FIGURE 4.3 Case 1

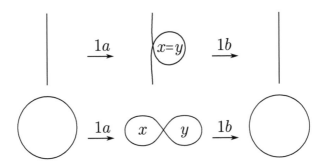

FIGURE 4.4 Case 1-(i)

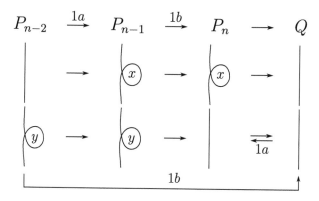

FIGURE 4.5 Case 1-(ii)

pair $(1a)(1b)$, as shown in Figure 4.5. Thus, the b-move (nth move) can be eliminated (Case 1-(i)) or moved to the position of the $(n-1)$th move (Case 1-(ii)).

• Case 2. We can draw two successive local moves $(2a)(1b)$, as shown in Figure 4.6. Denote by x (resp. y) a disk with a boundary generated (resp. eliminated) by $2a$ (resp. $1b$).

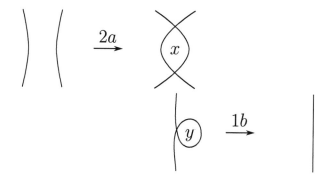

FIGURE 4.6 Case 2

• Case 2-(i) where $\partial x \cap \partial y \neq \emptyset$. In this case, there can be only one situation and we can eliminate the b-move as shown in Figure 4.7.
• Case 2-(ii) where $\partial x \cap \partial y = \emptyset$. In this case, as shown in Figure 4.8, the nth move can be moved to the position of the $(n-1)$th move by finding a bypass $(1b)(2a)$. The b-move (nth move) can be eliminated or moved to the position of the $(n-1)$th move.
• Case 3. We can draw two successive local moves $(1a)(2b)$, as shown in Figure 4.9. Denote by x (resp. y) a disk with a boundary generated (resp. eliminated) by $1a$ (resp. $2b$).

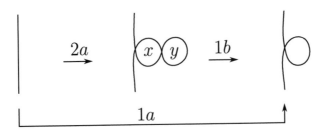

FIGURE 4.7 Case 2-(i)

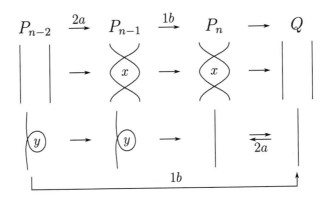

FIGURE 4.8 Case 2-(ii)

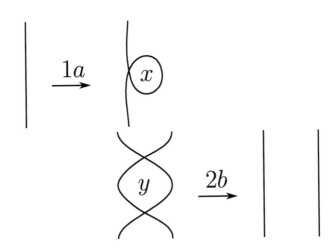

FIGURE 4.9 Case 3

• Case 3-(i) where $\partial x \cap \partial y = \emptyset$. In this case, we only have the situation as shown in Figure 4.10. The two successive local moves $(1a)(2b)$ can be replaced with $1b$.

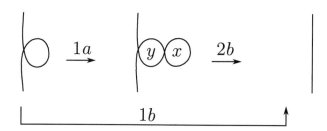

FIGURE 4.10   Case 3-(i)

• Case 3-(ii) where $\partial x \cap \partial y = \emptyset$. In this case, for a successive local move $(1a)(2b)$, depending on the pair, we can find a bypass $(2b)(1a)$ instead of the pair $(1a)(2b)$, as shown in Figure 4.11. Thus, the b-move (nth move) can be

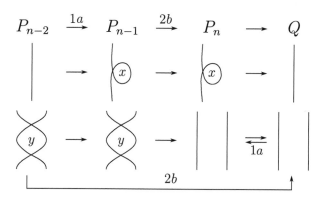

FIGURE 4.11   Case 3-(ii)

eliminated or moved to the position of the $(n-1)$th move (Case 3-(ii)).
• Case 4. We can draw two successive local moves (2a)(2b), as shown in Figure 4.12. Denote by x (resp. y) a disk with a boundary generated (resp. eliminated) by (2a) (resp. (2b)).
• Case 4-(i) where $\partial x \cap \partial y \neq \emptyset$. In this case, we have two situations, as shown in Figure 4.13. For each situation, two successive local moves (2a)(2b) can be replaced by $\emptyset$.
• Case 4-(ii) where $\partial x \cap \partial y = \emptyset$. In this case, for two successive local moves (2a)(2b), depending on the pair, we can find a bypass (2b)(2a) instead of the pair (2a)(2b), as shown in Figure 4.14. Thus, the b-move (nth move) can be eliminated or moved to the position of the $(n-1)$th move (Case 4-(ii)).

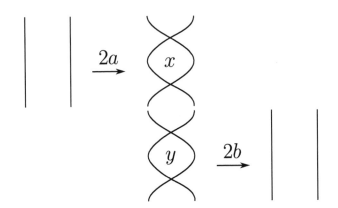

FIGURE 4.12 Case 4

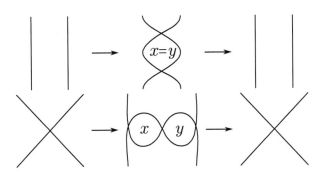

FIGURE 4.13 Case 4-(i)

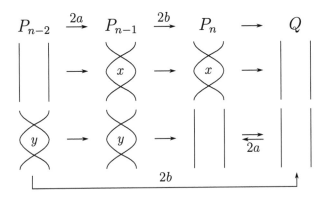

FIGURE 4.14 Case 4-(ii)

As a result, for every case, we can find a bypass removing b-move or shift b-move to the previous one. Here, we call the elimination and a shift *b-reduction*. We can repeat b-reduction and the b-move b we focus on can be move to the first move at starting point from P_0 if the b-move remains under these b-reductions. However, P_0 has no 1- or 2-gons. Thus, this b should be removed somewhere under repeating b-reductions. We repeat the discussion: we apply b-reduction if we encounter the first appearance of a b-move from the starting. Finally, we remove all the b-move in the result of the sequence. It completes the proof of Lemma 4.1.

By the structure of the proof of Lemma 4.1, we can easily obtain the following Lemma 4.2 and Theorem 4.3 because it is sufficient to consider only Case 1 or Case 4 from the proof of Lemma 4.1.

Lemma 4.2 *Let P_0 be a knot projection with no 1-gons (resp. 2-gons). If two knot projections P_0 and P are related by a finite sequence generated by RI (resp. RII), then we have a finite sequence generated by 1a (resp. 2a) from P_0 to P.*

If two knot projections P and P' are related by a finite sequence generated by RI (resp. RII), we write $P \overset{RI}{\sim} P'$ (resp. $P \overset{RII}{\sim} P'$).

Theorem 4.3 ([4]) *Let P and P' be knot projections. Let P^{1r} (resp. P^{2r}) be a knot projection with no 1-gons (resp. 2-gons) such that P^{1r} (resp. P^{2r}) is obtained through any finite sequence generated by (1b) (resp. (2b)) by eliminating the 1-gons (resp. 2-gons).*

$$P \overset{RI}{\sim} P' \Leftrightarrow P^{1r} = P'^{1r}.$$

$$P \overset{RII}{\sim} P' \Leftrightarrow P^{2r} = P'^{2r}.$$

4.3 CLASSIFICATION THEOREM UNDER RI AND strong RII OR RI AND weak RII

In this section, we consider the complete classification of knot projections under RI and strong RII or RI and weak RII. This result was first obtained by Ito–Takimura [5]. Strong RII (resp. weak RII) is a type of RII such that if the RII that increases the number of double points is applied, the 2-gon obtained is a *strong* (resp. *weak*) 2-gons, as shown in Figure 4.15. We introduce the result obtained by Ito–Takimura [5] as Theorem 4.4 and Theorem 4.5. To state Theorem 4.4 and Theorem 4.5, we first provide some definitions.

Definition 4.3 For two knot projections P and P', we write $P \overset{RI,\ strong\ RII}{\sim} P'$ (resp. $P \overset{RI,\ weak\ RII}{\sim} P'$) if there exists a finite sequence generated by $\overline{\Delta}$, RI, and strong RII (resp. weak RII) between P and P'.

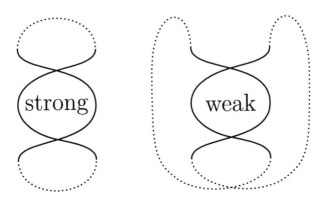

FIGURE 4.15 Strong (left) and weak (right) 2-gons

Definition 4.4 Let 1*b* (resp. 2*b*) be RI (resp. RII) decreasing the number of double points. If 2*b* belongs to a single strong RII, we denote it as *s*2*b* (resp. *w*2*b*). For a knot projection *P*, let P^{sr} (resp. P^{wr}) be a knot projection with no 1-gons and strong 2-gons (resp. weak 2-gons) obtained by a finite sequence generated by 1*b* and *s*2*b* (resp. *w*2*b*).

Theorem 4.4 ([5]) *The following two conditions (1) and (2) are equivalent.*
(1) Knot projections P and P′ are related by a finite sequence generated by RI and strong RII.
(2) Knot projections P^{sr} and P'^{sr} are sphere isotopic.

Proof 4.1 We can repeat the same argument using the proof of Lemma 4.1 for the proof of Lemma 4.3 (below) except for some 2-gons being restricted to strong 2-gons. The following table indicates which case is restricted to the strong type; this can be used to obtain the proof.

Case 1	same proof
Case 2-(i)	same proof
Case 2-(ii)	argument restricted to strong type
Case 3-(i)	same proof
Case 3-(ii)	argument restricted to strong type
Case 4	argument restricted to strong type

Lemma 4.3 ([5]) *Let P_0 and P be knot projections such that $P_0 = P_0{}^{sr}$ and* $P \overset{RI,\ \text{strong RII}}{\sim} P_0$. *Then, there exists a finite sequence generated by 1a and s2a from P_0 to P.*

Theorem 4.5 ([5]) *The following two conditions (1) and (2) are equivalent.*
(1) Knot projections P and P′ are related by a finite sequence generated by RI and weak RII.
(2) Knot projections P^{wr} and P'^{wr} are sphere isotopic.

Proof 4.2 By applying similar argument as that shown in the proof of Lemma 4.3, we can obtain Lemma 4.4 for the case of weak 2-gons; here, we show the cases that need to be restricted to the weak type. The following table can be used to obtain the proof.

Case 1	same proof
Case 2-(i)	weak 2-gon not to be considered for the proof
Case 2-(ii)	argument restricted to weak type
Case 3-(i)	weak 2-gon not to be considered for the proof
Case 3-(ii)	
Case 4	argument restricted to weak type

Lemma 4.4 ([5]) *Let P_0 and P be knot projections such that $P_0 = P_0{}^{wr}$ and $P_0 \overset{RI, \text{ weak } RII}{\sim} P$. Then, there exists a finite sequence generated by $1a$ and $w2a$ from P_0 and P.*

4.4 CIRCLE NUMBERS FOR CLASSIFICATION UNDER RI AND strong RII

Averaged invariant (see Chapter 7) is a famous integer-valued additive invariant under RI and strong RII (for definitions of *additivity* with respect to connected sums, see Chapter 6). Here, we obtain a property to calculate the averaged invariant $a(P)$ for a knot projection P as follows.

Property 4.1 $a(P)$ can be calculated by determining a finite sequence from P to a trivial knot projection. Similarly to RII, a single RIII is decomposed into two types by the connections of the three arcs of 3-gon, as shown in Figure 4.16. For each of the moves, i.e., RI, strong RII, weak RII, strong RIII, and weak RIII, we define an *a*-move (see Figure 4.16). RI that increases the number of double points is denoted by $1a$. Similarly, strong RII (resp. weak RII) increasing the number of double points is denoted by $s2a$ (resp. $w2a$). For strong RIII (resp. weak RIII), $s3a$ (rep. $w3a$) is defined by Figure 4.16. Every move $*a$ is called an *a*-move. Each inverse move is called a *b*-move, denoted by move $*b$, where a is replaced by b. The averaged invariant $a(P)$ has the following property. By Reidemeister's theorem for knot projection, we can compute $a(P)$ in an inductive manner.

1. $a(\bigcirc) = 0$ for a trivial knot projection $\bigcirc$.

2. RI and strong RII do not change $a(P)$.

3. A single $w2a$ changes $a(P)$ by $+1$.

4. A single $s3a$ changes $a(P)$ by $+1$.

5. A single $w3a$ changes $a(P)$ by -1.

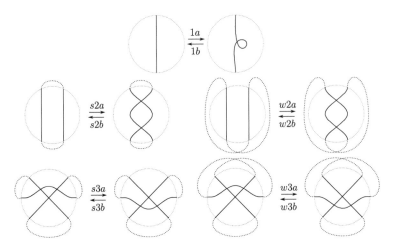

FIGURE 4.16 $1a$, $1b$, $s3a$, $s3b$, $w3a$, and $w3b$

Ito and Takimura [5] introduced another integer-valued additive invariant, called *circle number*, of knot projections under RI and strong RII. To the best of our knowledge, except for the averaged invariant and circle number, an invariant that is integer-valued, additive, and invariant under RI and strong RII is not known. The circle number is defined as follows. We give an orientation o to a knot projection P. Every double point of the knot projection (P, o) with orientation o is replaced as shown in Figure 4.17. Then, we have an ar-

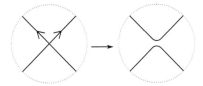

FIGURE 4.17 Non-Seifert splice

rangement of the circles on a 2-sphere, where there is no double point. The arrangement of circles does not depend on orientation o. If an opposite orientation, $\bar{o}$, is chosen and a non-Seifert splice is applied to every double point of $(P, \bar{o})$, we obtain the same arrangement of circles. This is because switching orientations, $o \to \bar{o}$, does not change the result in Figure 4.17 for a 2-sphere. Thus, an arrangement is uniquely determined by a knot projection P. The arrangement of circles of P is called a *circle arrangement* and is denoted by $\tau(P)$. The number of circles in $\tau(P)$ is called a *circle number* and is denoted by $|\tau(P)|$. We obtain some examples in Figs. 4.18 and 4.19.

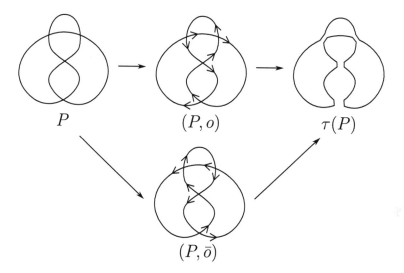

FIGURE 4.18 Circle arrangements without orientations

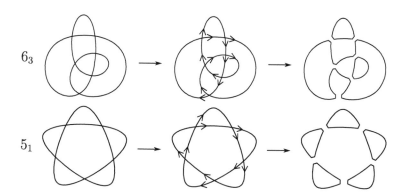

FIGURE 4.19 Two circle arrangements

Theorem 4.6 ([5]) *Circle arrangements and circle numbers are invariant under RI and strong RII.*

The averaged invariant $a(P)$ takes values $a(6_3) = a(5_1) = 2$. However, $|\tau(6_3)| = 3 \neq 5 = |\tau(5_1)|$. Thus, $\tau(P)$ is independent of $a(P)$. On the other hand, for a $(3, 4)$-torus knot projection $T_{3,4}$, $\tau(T_{3,4}) = \bigcirc = \tau(\bigcirc)$. However, $a(T_{3,4}) = 4$ (see Figure 4.20). As shown in Figure 4.22, circle arrangement $\tau(P)$ is strictly a stronger invariant than circle number $|\tau(P)|$ under RI and strong RII [5, Proposition 1].

Ito and Takimura obtained the complete classification theorem for knot projections under RI and strong RII, or RI and weak RII [5]. Essentially, the complete invariant is $\mathcal{C} \to \mathcal{C}; P \mapsto P^{sr}$ (resp. P^{wr}) for RI and strong RII (resp. weak RII), where $\mathcal{C}$ is the set of the knot projections on a sphere. On the other hand, $\tau(P)$ can be used to obtain the characteristics of some knot projections. We introduce some properties shown in [5].

Theorem 4.7 ([5]) ● *Every circle number is an odd integer.*

● *For a knot projection P, P_o denotes P with an arbitrary orientation. If P_o has a sufficiently small disk containing $2n$-gon $(n \geq 1)$ satisfying the following condition, then $|\tau(P)| \geq 3$.*

Condition: Suppose that we obtain an arbitrary orientation to a knot projection. Any two neighbored arcs have opposite orientations, as shown in Figure 4.23.

Corollary 4.2 *A knot projection with $|\tau(P)| = 1$ has no weak 2-gon.*

Theorem 4.8 ([5]) *Let P be a knot projection and let $\bigcirc$ be a trivial knot projection. The following conditions are mutually equivalent.*
(1) $P \overset{RI,\ strong\ RII}{\sim} \bigcirc$ *and* $P \overset{RI,\ weak\ RII}{\sim} \bigcirc$.
(2) P *can be obtained from* $\bigcirc$ *by a finite sequence generated by* $1a$.

Proof 4.3 It is sufficient to check (1) $\Rightarrow$ (2). From condition (1), $|\tau(P)| = \tau(\bigcirc) = 1$ and there exists a finite sequence generated by $1a$ and $w2a$ from $\bigcirc$ to P (Lemma 4.4). Now we assume that there exists at least one $w2a$ in the sequence $(\star)$. Then, when going from P to $\bigcirc$, there exists $w2b$ that is firstly encountered. We denote the knot projection just before the encountered $w2b$ as P'. Thus, P and P' are related by a finite sequence generated by RI such that $|\tau(P)| = |\tau(P')| = 1$. The knot projection P' has at least one weak 2-gon. However, Corollary 4.2 implies that this is a contradiction. Hence, $(\star)$ is false. Thus, there exists a finite sequence generated by RI from $\bigcirc$ to P. Further, by Theorem 4.3, there exists a finite sequence generated by $1a$ from $\bigcirc$ to P.

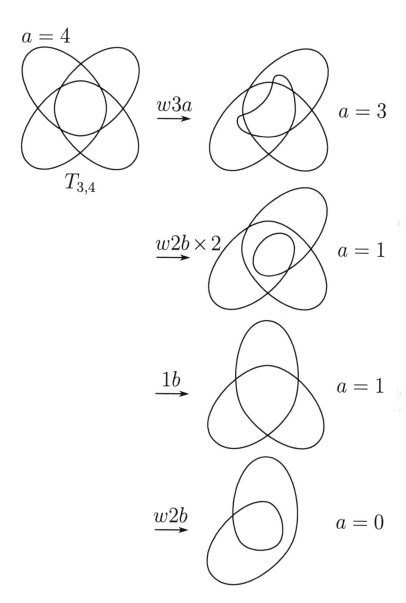

FIGURE 4.20 Value $a(T_{3,4})$

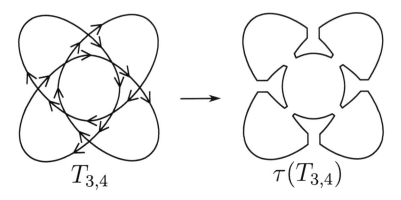

$$T_{3,4} \qquad \tau(T_{3,4})$$

FIGURE 4.21 Value $\tau(T_{3,4})$

4.5 EFFECTIVE APPLICATIONS OF THE CIRCLE NUMBER

In this section, we suppose that every knot projection has at least one double point.

Shimizu [9] introduced the notion of reductivity with respect to a geometric problem as follows:

Problem 4.1 *Let symbols A, B, C, and D represent the types of 3-gons, as shown in Figure 4.24. Is it true that any knot projection with no 1-gons or 2-gons has at least one 3-gon of type A, B, or C?*

If the answer to Problem 4.1 is *yes*, then Shimizu's reductivity r satisfies $r \leq 3$ for any knot projection. We introduce Shimizu's second question (Question 4.1) as follows:

Question 4.1 ([9]) *Is it true that $r(P) \leq 3$ for every knot projection P?*

Now, we introduce the definition of reductivity r.

Definition 4.5 (reducible knot projection) Let P be a knot projection. The replacement of a sufficiently small disk of a double point with another disk, as shown in Figure 4.25, is called a *splice*. If we give P an arbitrary orientation, we obtain two types of splices; one is called a *Seifert splice* (Figure 4.26, upper) and the other is called a *non-Seifert splice* (Figure 4.26, lower). By definition, both Seifert splice and non-Seifert splice do not depend on the choice of orientation of P.

Suppose that we apply a Seifert splice to a double point d, then we have two knot projections P_1 and P_2 such that there is no double point between P_1 and P_2. Then, double point d is called a *nugatory* double point. If a knot projection P has a nugatory double point, then P is called a *reducible* knot

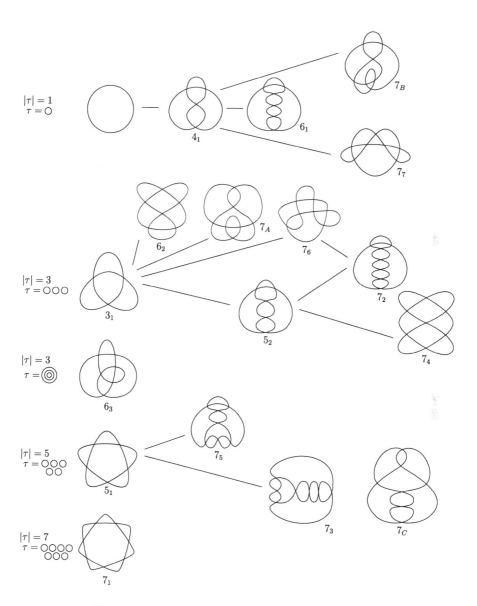

FIGURE 4.22 Circle arrangements. A segment is drawn if a path consisting of finitely many RIs and a single strong RII is found

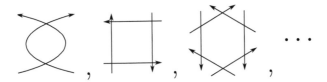

FIGURE 4.23 $2n$-gons $(n = 1, 2, 3, \ldots)$

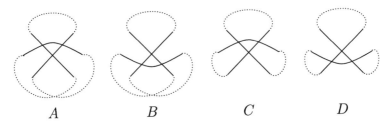

| A | B | C | D |

FIGURE 4.24 3-gons of type A, B, C, and D from left to right.

projection. If P has no nugatory double point, then P is called a *reduced* knot projection.

Definition 4.6 ([9]) Let P be a knot projection. The reductivity $r(P)$ is the minimum number of non-Seifert splices applied recursively to obtain a reducible knot projection from P.

Question 4.1 is still an open question. In general, it is not easy to calculate r. The circle number is useful for the computation of r (Theorem 4.9).

Theorem 4.9 ([7]) *Let P be a reduced knot projection.*

$$|\tau(P)| = 1 \Longrightarrow 2 \le r(P).$$

To prove Theorem 4.9, we use Lemma 4.5.

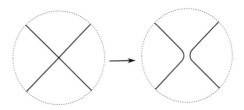

FIGURE 4.25 Splice

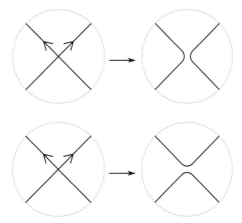

FIGURE 4.26 Seifert splice (upper) and non-Seifert splice (lower)

Lemma 4.5 (**[6]**) *Let P be a knot projection. $r(P) = 1$ if and only if there exists a circle intersecting P at just two double points, as shown in Figure 4.27.*

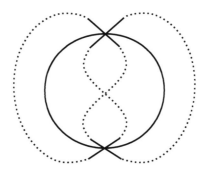

FIGURE 4.27 Knot projection P with a circle that divides S^2 into two disks and intersects at just two double points of P.

Proof of Lemma 4.5.
(*If part*) Apply a non-Seifert splice at one of the double points on the circle.
(*Only if part*) Let P be a knot projection. If P is a reducible, then there exists a disk, a *reducible disk*, containing a part of P such that its boundary, i.e., a circle, intersects P at a single double point only (left figure of Figure 4.28).

If $r(P) = 1$, then there exists a non-Seifert splice, from a double point to two sub-curves, such that one sub-curve is inside the reducible disk and the other is outside (right of Figure 4.28). In the aforementioned situation, there

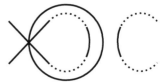

FIGURE 4.28 A part (specified by a single circle) with a reducible knot projection

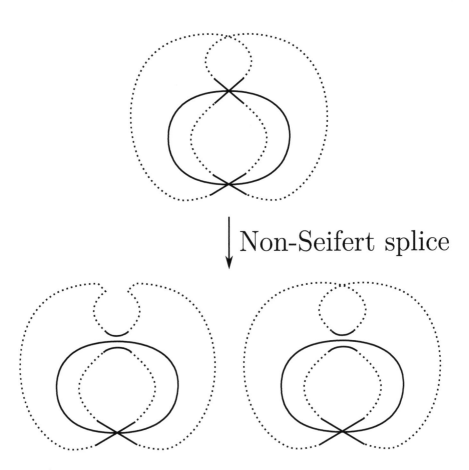

Non-Seifert splice

FIGURE 4.29 Possibility of producing a reducible knot projection by a single non-Seifert splice

exist exactly two possibilities of connections of the two sub-curves obtained by applying the non-Seifert splice, which are indicated by dotted sub-curves in Figure 4.29 (bottom line). Suppose that these two possibilities are the results obtained. Then, we apply an inverse operation of the non-Seifert splice at the position indicated by solid arcs, as shown in the bottom line of Figure 4.29.

Therefore, if $r(P) = 1$, then P has a circle intersection just at two double points of P, as shown in the upper line of Figure 4.29, equivalently, Figure 4.27.

□

Now we will prove Theorem 4.9. If we apply non-Seifert splices at two double points on the circle as mentioned in the statement of Lemma 4.5, then we have $|\tau(P)| \geq 2$. That is, for a reduced knot projection P, we have

$$r(P) = 1 \Longrightarrow |\tau(P)| \geq 2.$$

Then, we have the contraposition. For a reduced knot projection P,

$$|\tau(P)| = 1 \Longrightarrow r(P) \geq 2.$$

This completes the proof.

□

4.6 FURTHER TOPICS

Alternatively, Taniyama introduced another type of reductivity, t (private communication). In this section, we suppose that every knot projection has at least one double point.

For Taniyama's reductivity t, we can state the following:
(1) If the answer to Problem 4.1 is yes, then $t(P) \leq 3$ ($\forall P$).

Thus, it is worth estimating t to obtain a negative answer to Problem 4.1. Note that it is easier to obtain a positive answer to (1) than to Question 4.1 because $t \leq r$. Taniyama's reductivity t is defined as follows.

Definition 4.7 ([7]) Let P be a knot projection. The reductivity $t(P)$ is the minimum number of splices applied simultaneously to obtain a reducible knot projection from P.

By definition, $t \leq r$. Surprisingly, we have not found an example for $t(P) = 3$. Thus, we have the following question:

Question 4.2 *Is it true that $t(P) \leq 2$ for every knot projection P?*

The proposition as shown in [7] shows:

Proposition 4.1 ([7]) *For every knot projection P,*

$$t(P) = 1 \Longleftrightarrow r(P) = 1.$$

Thus, we can estimate t by using circle number τ:

Proposition 4.2 ([7]) *Let P be a reduced knot projection.*

$$|\tau(P)| = 1 \Longrightarrow 2 \le t(P)(\le r(P)).$$

Further, the necessary and sufficient condition that for a knot projection P, $t(P) = 2$, is also shown in [7]. In order to obtain the condition $t(P) = 2$, [7] introduced new reductivities y and i ([7]).

Definition 4.8 ([7]) The reductivity $y(P)$ is the minimum number of non-Seifert splices applied simultaneously to obtain a reducible knot projection from P.

Definition 4.9 ([7]) The reductivity $i(P)$ is the minimum number of Seifert splices applied simultaneously to obtain a reducible knot projection from P.

We would like to conclude this chapter with the homotopies obtained by (RI, RII), (RI, strong RII), and (RI, weak RII) here. For further details on reductivities, see Chapter 9.

4.7 OPEN PROBLEMS AND EXERCISE

1. (open) Determine the knot projections, each of whose circle number is one. For example, the following sequence shows knot projections, each of whose circle number is one (Figure 4.30).

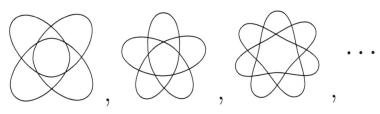

FIGURE 4.30 A sequence of knot projections

2. (open) Find a new integer-valued additive invariant under RI and strong RII.

3. (open) Find a new integer-valued additive invariant under RI and weak RII.

4. (open) Solve Problem 4.1.

5. (open) Solve Question 4.1.

6. (open) Provide a positive or negative answer to Question 4.1.

7. Find a knot projection such that $t(P) \ne r(P)$.

Bibliography

[1] V. I. Arnold, Topological invariants of plane curves and caustics. Dean Jacqueline B. Lewis Memorial Lectures presented as Rutgers University, New Brunswick, New Jersey. University Lecture Series, 5. *American Mathematical Society, Providence, RI, 1994*. viii+60 pp.

[2] R. Fenn, P. Taylor, Introducing doodles, *Topological of low-dimensional manifolds (Proc. Second Sussex Conf., Chelwood Gate, 1977)*, pp. 37–43, Lecture Notes in Math., **722**, *Springer, Berlin, 1979*.

[3] N. Ito and A. Shimizu, The half-twisted splice operations on knot projections, *J. Knot Theory Ramifications* **21** (2012), 1250112, 10pp.

[4] N. Ito and Y. Takimura, (1, 2) and weak (1, 3) homotopies on knot projections, *J. Knot Theory Ramifications* **22** (2013), 1350085, 14pp.

[5] N. Ito and Y. Takimura, Strong and weak (1, 2) homotopies on knot projections, to appear in *Kobe J. Math.* To be published.

[6] N. Ito and Y. Takimura, Triple chords and strong (1, 2) homotopy, to appear in *J. Math. Soc. Japan.* **68** (2016), 637–651.

[7] N. Ito and Y. Takimura, Knot projections with reductivity two, *Topology Appl.* 193 (2015), 290–301.

[8] M. Khovanov, Doodle groups, *Trans. Amer. Math. Soc.* **349** (1997), 2297–2315.

[9] A. Shimizu, The reductivity of spherical curves, *Topology Appl.* **196** (2015), part B, 860–867.

CHAPTER 5

Classification by RI and strong or weak RIII (1996–2015)

CONTENTS

5.1 An example by Hagge and Yazinski 53
5.2 Viro's strong and Weak RIII 55
5.3 Which knot projections trivialize under RI and weak RIII? 57
5.4 Which knot projections trivialize under RI and
     Strong RIII? 61
5.5 Open problem and Exercises 70

I f RIII is forbidden, a knot projection will be non-trivial; this has been shown by Hagge–Yazinski [1] (Chapter 7). However, it is still unknown that which knot projection is trivialized by a finite sequence generated by RI and RIII. By contrast, the classification of knot projections under the equivalence relation generated by RII and RIII has been completed (Chapter 3). We have also learned that the classification of knot projections under the equivalence relation generated by RI and RII is complete (Chapter 4).

In order to address the classification problem, as the first step, we consider two types of RIII, i.e., strong and weak moves, and consider the pairs (RI, strong RIII) and (RI, weak RIII).

5.1 AN EXAMPLE BY HAGGE AND YAZINSKI

Throughout this book, if two knot projections P and P' are related by a finite sequence generated by $\overline{\Delta}$, then P and P' are considered to be equivalent and we write $P = P'$. Thus, in the rest of this book, we often omit the use of $\overline{\Delta}$; e.g., a finite sequence generated by RI and RIII indicates a finite sequence generated by $\overline{\Delta}$, RI, and RIII.

In Chapters 3 and 4, we provided classifications in terms of the pair (RII,

RⅢ) or (RI, RⅡ). The following problem was obtained from the study of Östlund ([5], 2001).

Problem 5.1 *For a given knot projection, can we find a finite sequence generated by RI and RⅢ such that the knot projection is related to a trivial knot projection?*

The example P_{HY} by Hagge–Yazinski is shown in Figure 5.1. There is no finite

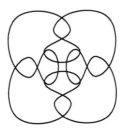

FIGURE 5.1 Non-trivial knot projection P_{HY} under RI and RⅢ

sequence generated by RI and RⅢ between knot projection P_{HY}, as shown in Figure 5.1, and a trivial knot projection. This non-triviality will be shown in Chapter 7.

Problem 5.2 *There exist the following open problems.*
(1) Determining which knot projection can be trivialized by RI and RⅢ.
(2) Determining a maximum set S consisting of knot projections such that any two elements are related by a finite sequence generated by RI and RⅢ.

A maximum set S_M is defined as the set satisfying $S \subset S_M$, for any set S that holds a certain condition. Here, the condition is that any two elements of S are related by a finite sequence generated by RI and RⅢ. In mathematics, such a set, S_M, is called an *equivalence class*. We notice that Problem 5.2 (2) contains (1), i.e., it is the case when the set S contains a trivial knot projection. In this situation, we would like to specify the equivalence class S_M containing a trivial knot projection $\bigcirc$, e.g., $[\bigcirc]$. Thus, in general, an equivalence class containing a knot projection P is denoted by $[P]$. Problem 5.2 can be rewritten as follows.

Problem 5.3 *(1) Determine $[\bigcirc]$.*
(2) Determine $[P]$ for a knot projection P.

It is surprising that at least one equivalence class has not yet been determined.
Before we proceed to the next section, we will provide the definition of an *equivalence class* in a general situation for the readers who are interested.
Let S be a set. A subset R of $S \times S = \{(s, s') \mid s,\ s' \in S\}$ is called a *relation*. If $(s, s') \in R$, we write sRs'. If R satisfies the following three properties, we call it an *equivalence relation*.

1. xRx for any $x \in \mathcal{S}$,

2. $xRy \Rightarrow yRx$ for any $x, y \in \mathcal{S}$,

3. xRy and $yRz \Rightarrow xRz$ for any $x, y, z \in \mathcal{S}$.

We often denote an equivalence relation R by $\sim$. Take an arbitrary $x \in \mathcal{S}$ and consider the set $C_x = \{y \mid y \sim x\}$. By considering various values of x, we obtain the following properties:

1. $C_x \cap C_y = \emptyset$.

2. We can choose a set $\Lambda \subset \mathcal{S}$, where $\mathcal{S} = \cup_{x \in \Lambda} C_x$.

Every C_x is called an *equivalence class* and is denoted by $[x]$.

In particular, set $\mathcal{S} = \{\text{knot projections}\}$. $P \overset{\text{RI, RIII}}{\sim} P'$ is defined as follows: $P \overset{\text{RI, RIII}}{\sim} P'$ if and only if there exists a finite sequence generated by $\overline{\Delta}$, RI, and RIII. The relation $\overset{\text{RI, RIII}}{\sim}$ gives an equivalence relation.

5.2 VIRO'S STRONG AND Weak RIII

Viro specified strong and weak RIII to study the generalization (quantization) of Arnold invariant J^-, which is an invariant of knot projections on a plane under weak RII and weak RIII ([6], 1996). We provide a short introduction of Viro's quantization in Chapter 11 and here, we introduce the recent works of Ito–Takimura [2] and Ito–Takimura–Taniyama [3], that are concerned with strong RIII and weak RIII.

We know that each Reidemeister move is a local move; that is, there exists a sufficiently small disk and we can represent it as the replacement of the sub-curves in the disk. For example, a single RIII can be represented by a replacement in the disk. The boundary of the disk and the sub-curves intersect at six distinct points. Considering a knot projection, the six points can be connected in exactly four ways. This is because these four ways of connecting the six points correspond to all the possibilities of 3-gons appearing in a knot projection. Next, we notice that there are exactly two pairs, each of which can be connected by a single RIII, as shown in Figure 5.2. One is called a strong RIII and the other is called a weak RIII.

Definition 5.1 A strong RIII and weak RIII are defined as shown in Figure 5.2. We can detect which RIII is strong (or weak) by using an arbitrary chosen orientation (Figure 5.3). RI, strong RII, weak RII, strong RIII, and weak RIII are called *Reidemeister moves* or *refined Reidemeister moves*.

Definition 5.2 Let x and y be Reidemeister moves. For two knot projections P and P', denoted by $P \overset{x,y}{\sim} P'$ (resp. $P \overset{x}{\sim}$), if P can be related to P' by a finite sequence generated by $\overline{\Delta}$, x, and y (resp. $\overline{\Delta}$ and x). If $P \overset{x,y}{\sim} P'$ (resp. $P \overset{x}{\sim} P'$) is not possible, we write $P \overset{x,y}{\nsim} P'$ (resp. $P \overset{x}{\nsim} P'$).

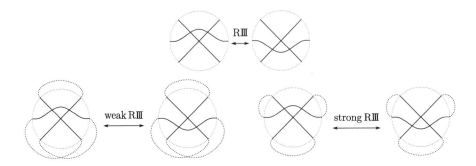

FIGURE 5.2 RⅢ (first line), strong RⅢ, and weak RⅢ (second line). Dotted sub-curves indicate the fragment connections

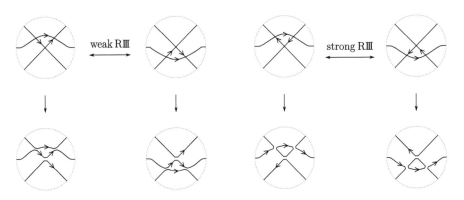

FIGURE 5.3 Characterization of strong and weak RⅢ

5.3 WHICH KNOT PROJECTIONS TRIVIALIZE UNDER RI AND weak RIII?

Theorem 5.1 ([2]) *Let* $\bigcirc$ *be a trivial knot projection and* P *be a knot projection. A single RI decreasing double points is denoted by 1b.*

$$P \overset{RI, \ weak \ RIII}{\sim} \bigcirc \Leftrightarrow P \overset{1b}{\sim} \bigcirc.$$

Here, we introduce the first proof of Theorem 5.2 through knots. In Chapter 6, we introduce another proof through a notion of an *invariant* under RI and weak RIII.

Proof 5.1 Let us call each double point of a knot diagram (resp. knot projection) a *crossing* (resp. a double point). For a given knot projection, we replace each double point with a crossing, called a *positive crossing*, as shown in Figure 5.4. The replacement is called a *positive resolution*. This resolution

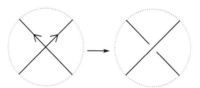

FIGURE 5.4 Positive resolution

gives rise to the following map:

$$f : \{\text{knot projections}\} \to \{\text{knot diagrams}\}.$$

Here, let us call an equivalence relation generated by RI and weak RIII, *weak* (1, 3) *homotopy.*

Consider a weak (1, 3) homotopy class. The images of a single RI and a weak RIII obtained by f are shown in Figure 5.5, and are denoted by $f(\text{RI})$ and $f(\text{weak RIII})$, respectively. It can be easily observed that $f(\text{RI})$ (resp. $f(\text{weak RIII})$) is a first (resp. a kind of a third) Reidemeister move of knot diagrams.

Thus, the following two maps f_1 and f_2 are well defined.

$$\{\text{knot projections} \, / \, \text{weak} \, (1, 3) \, \text{homotopy}\} \overset{f_1}{\to} \{\text{knot diagrams} \, / \, \text{RI, weak RIII}\}$$

$$\overset{f_2}{\to} \{\text{knot diagrams} \, / \, \text{RI, RII, RIII}\}$$

$$= \{\text{knots}\}.$$

$$(5.1)$$

We set a map $[f] = f_2 \circ f_1$. Here, we recall the following fact of knot theory (we will describe it later in Chapter 6 for the readers who are not familiar

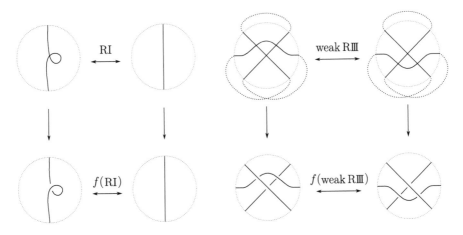

FIGURE 5.5 $f(\mathrm{RI})$ and $f(weak\ \mathrm{RIII})$

with this). A knot is called a *positive knot K* if there exists a knot diagram of K corresponding to a knot satisfying the condition that all the crossings are positive crossings; such a knot diagram is called a *positive knot diagram*. Recall that $\mathcal{RI}$ is a local replacement as shown in Figure 2.2.

Fact 5.1 *If the knot of a positive knot diagram is a trivial knot, then the positive knot diagram can be related to a trivial knot diagram by a finite sequence generated by $\mathcal{RI}$ decreasing the number of double points.*

For a knot projection P, a knot diagram $[f](P)$ can be related to a trivial knot projection by using only $\mathcal{RI}$ and decreasing the number of double points. Thus, omitting the over-/under- information of $[f](P)$, we have the claim.

We state Proposition 5.1 here. By using Proposition 5.1, we can easily show that the equivalence classes under weak $(1, 3)$ homotopy are non-trivial.

Proposition 5.1 ([2]) *If there is a knot invariant I, then $I \circ [f]$ is a weak $(1, 3)$ homotopy invariant of the knot projections.*

Proof 5.2 Figure 5.5 shows the claim. Here, recall that if two knot diagrams represent a knot isotopy class, then they are related by a finite sequence generated by $\mathcal{RI}$, $\mathcal{RII}$, and $\mathcal{RIII}$ (cf. Figure 2.2).

Corollary 5.1 *Let P_{HY} be the knot projection as shown in Figure 5.1.*
$$P_{HY} \overset{RI,\ weak\ RIII}{\sim} \bigcirc.$$

Proof 5.3 For example, if we consider a 3-*colorable* function, Col, as a knot invariant I (for Col, see Remark 5.1), then a knot with representative $f(P_{HY})$ is not a trivial knot. This implies the claim.

Remark 5.1 The function Col is defined as follows. For a knot diagram, every crossing has over and under information; thus, every crossing consists of two paths, i.e., the over and the under path. In the neighborhood of a crossing of a knot diagram, the under path is cut at the crossing and divided into two parts. For every crossing, we apply to cutting the under path. Two boundaries obtained from this cutting are called edge-points. The connected part between the edge-points in a knot diagram is called an *arc* for a knot diagram (i.e., note that the notion of arcs of a knot projection is different from that of knot diagrams, cf. Notation 4.1). A knot diagram consists of a finite number of arcs. Here, we consider the following two rules.

1. At a crossing, three arcs can have either three colors or one color.

2. A knot diagram should be colored with least two colors.

We obtain a colored knot diagram D by following the above rules. Then, we define the function Col as follows.

$$\mathrm{Col}(D) = \begin{cases} 1 & D \text{ is 3-colorable,} \\ 0 & \text{otherwise.} \end{cases}$$

Fact 5.2 (a well-known fact) *Let D be a diagram of knot K. $\mathrm{Col}(D)$ does not depend on the choice of D. In other words, Col is a knot invariant.*

Proof 5.4 We compare two local disks concerned with Reidemeister moves of three types. We need to show that if one color is fixed, then the color of the other diagram is also fixed. Now, we represent three colors by c_1, c_2, c_3 (see Figure 5.6).
● (Invariance under $\mathcal{RI}$) Observe the two endpoints on a circle for each of the two disks with respect to $\mathcal{RI}$. If we choose a color c_1 for one of the endpoints, the color for the other endpoint should also be c_1. This implies $\mathrm{Col}(D) = \mathrm{Col}(D')$ between two knot diagrams D and D' for a single $\mathcal{RI}$.
● (Invariance under $\mathcal{RII}$) Observe the four endpoints on the circle for each of the two disks with respect to $\mathcal{RII}$. The four endpoints have two strands; thus, the endpoints are divided into two pairs p_1 and p_2, where two endpoints of the same strand belong to the same pair. Let D and D' be two knot diagrams related by a single $\mathcal{RII}$. There are exactly two possibilities for the coloring, as shown in Figure 5.6. Then, there exists a one-to-one correspondence between D and D', which implies $\mathrm{Col}(D) = \mathrm{Col}(D')$.
● (Invariance under $\mathcal{RIII}$) Let D and D' be two knot diagrams related by a single $\mathcal{RIII}$. Observe the six endpoints on a circle for each of the two disks with respect to $\mathcal{RIII}$. There exist exactly five cases for coloring the disk corresponding to a single $\mathcal{RIII}$. For each case, if the coloring of D is determined, the coloring of D' is completely determined because the colors of all the six endpoints can be then fixed. This implies $\mathrm{Col}(D) = \mathrm{Col}(D')$.
  Thus, we show the claim.

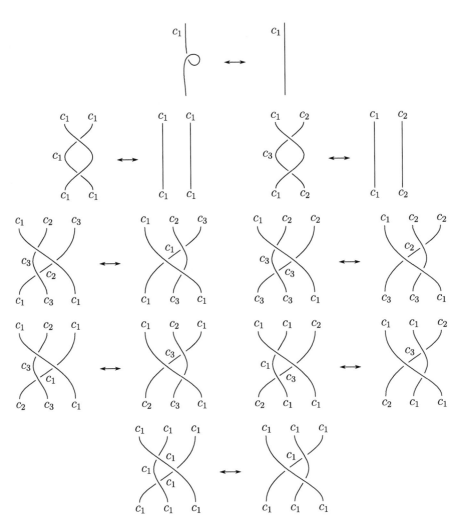

FIGURE 5.6 Reidemeister moves and 3-colorable knot diagrams

5.4 WHICH KNOT PROJECTIONS TRIVIALIZE UNDER RI AND Strong RⅢ?

In this section, to state Theorem 5.2, we define the following terminology.

Definition 5.3 A single strong RⅢ from the left figure to the right figure (resp. the right figure to the left figure) in Figure 5.7 is denoted by $s3a$ (resp. $s3b$). Let $1a$ be a single RI increasing the number of double points.

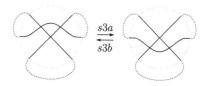

FIGURE 5.7 $s3a$ and $s3b$

Theorem 5.2 ([3]) *Let P be a knot projection and $\bigcirc$ be a trivial knot projection. $P \overset{RI,\ strong\ RⅢ}{\sim} \bigcirc$ if and only if there exists a finite sequence generated by $1a$ and $s3a$ from $\bigcirc$ to P.*

Proof 5.5 By definition, it is sufficient to check whether there exists a finite sequence generated by $1a$ and $s3a$ from $\bigcirc$ to a knot projection P if $\bigcirc \overset{RI,\ strong\ RⅢ}{\sim} P$.

Note that we cannot apply $1b$ or $s3b$ to a trivial knot projection $\bigcirc$. For a sequence of length n generated by $1a$ and $s3a$ from $\bigcirc$, we focus on the first appearance of $1b$ or $s3b$ at the $(n+1)$th position $(n \geq 1)$. Let u be a sequence of length $(n-1)$ generated by $1a$ and $s3a$ as shown below.

• Case 1: $u(1a)(1b)$. We have already considered this case in the proof of Lemma 4.1 in Chapter 4. Hence, we can replace $u(1a)(1b)$ with u or $u(1b)(1a)$.

• Case 2: $u(1a)(s3b)$. Let x be a 1-gon generated by $(1a)$ and y be a 3-gon eliminated by $(s3b)$ with respect to the last two local replacements of $u(1a)(s3b)$ (Figure 5.8). Below, we split the cases based on the conditions of the intersection of x and y.

• Case 2-(i): $\partial x \cap \partial y \neq \emptyset$.

It can be seen that there is no possibility of the existence of $(1a)(s3b)$ such that it satisfies the condition in the case.

• Case 2-(ii): $\partial x \cap \partial y = \emptyset$. In this case, because a knot projection satisfies the above condition, we can exchange the order of $(1a)$ and $(s3b)$, as shown in Figure 5.9.

• Case 3: $u(s3a)(1b)$. Let x be a 3-gon generated by $(s3a)$ and y be a 1-gon eliminated by $(1b)$ in the last two local replacements of $u(s3a)(1b)$ (Figure 5.10). Similarly to Case 2, we divide cases by the condition with respect to an intersection of x and y.

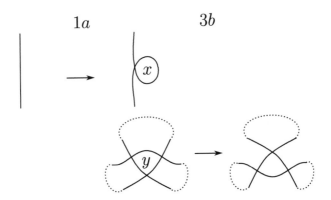

FIGURE 5.8 Case 2

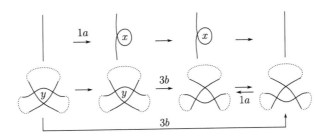

FIGURE 5.9 Case 2-(ii)

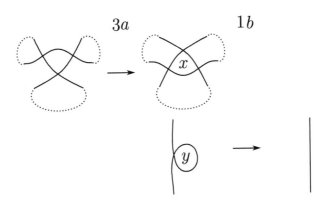

FIGURE 5.10 Case 3

• Case 3-(i): $\partial x \cap \partial y \neq \emptyset$. It can be seen that it is not possible for the observed last two local replacements $(s3a)(1b)$ to satisfy the condition.

• Case 3-(ii): $\partial x \cap \partial y = \emptyset$. In this case, we can exchange the two replacements $(s3a)$ and $(1b)$, as shown in Figure 5.11.

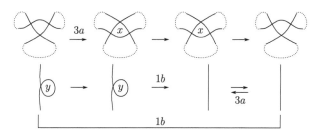

FIGURE 5.11   Case 3-(ii)

• Case 4: $u(s3a)(s3b)$. Let x be a 3-gon generated by $(s3a)$ and y be a 3-gon eliminated by $(s3b)$ in the last two local replacements of $u(s3a)(s3b)$ (Figure 5.12). Below, we divide the cases based on the condition of the intersection of x and y.

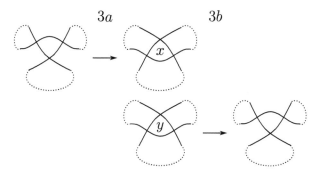

FIGURE 5.12   Case 4

• Case 4-(i): $\partial x \cap \partial y \neq \emptyset$. Note that a 3-gon in a knot projection consists of three arcs and a 3-gon has three double points. Because x and y are either eliminated or generated by a single strong RIII, x and y cannot share an arc. Therefore, there is a possibility that x and y may share m double points $(1 \leq m \leq 3)$. Figure 5.13 shows all the possibilities $(1 \leq m \leq 3)$ after $(s3a)$ and just before $(s3b)$.

At this point, we use Fact 5.3 (= Proposition 6.3) from Chapter 6.

Fact 5.3 (Proposition 6.3) *Let P be a knot projection. If $P \overset{RI, \ strong \ RIII}{\sim} \bigcirc$, then P cannot contain x and y $(1 \leq m \leq 2)$, as shown in Figure 5.13.*

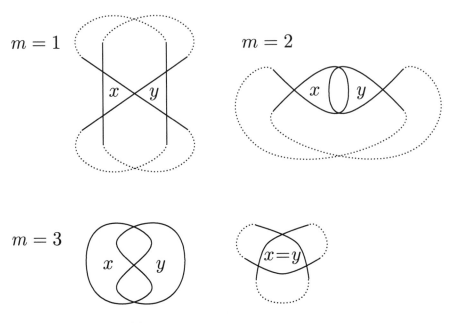

FIGURE 5.13 Case 4-(i)

Next, we show the case where $m = 3$. For each case, $u(s3a)(s3b)$ is replaced with u, as shown in Figure 5.14.

• Case 4-(ii): $\partial x \cap \partial y = \emptyset$. In this case, for a given sequence $u(s3a)(s3b)$, we can find another sequence $u(s3b)(s3a)$, as shown in Figure 5.15.

To summarize Case 1–Case 4, we show the following. Suppose that there exists a sequence $u(*a)(*b)$ from $\bigcirc$ to P, where u is generated by $(1a)$ and $(s3a)$, and $(*a) = (1a)$ or $(s3a)$, $(*b) = (1b)$ or $(s3b)$. Then, the sequence $u(*a)(*b)$ is replaced with u or $u(*b)(*a)$.

For a given sequence generated by RI and strong RIII from $\bigcirc$ to P, consider the case of the first appearance of a b-move, where the b-move is either $(1b)$ or $(s3b)$. If we encounter the first appearance of a b-move, say at nth move, the sequence is $u(*a)(*b)$ locally. From the above discussion, the b-move can be canceled or moved to the $(n-1)$th position. By repeating this discussion, we have two possibilities: (1) a b-move does not appear in the sequence u and (2) there is a b-move that is the first move in u. However, if a b-move is the first move from $\bigcirc$, then it is a contradiction. This is because for a trivial knot projection $\bigcirc$ and a knot projection Q, $\bigcirc \overset{1b}{\to} Q$ or $\bigcirc \overset{s3b}{\to} Q$ is not possible. Thus, only possibility (1) remains.

From the above discussion, we have a sequence from $\bigcirc$ to P for a given knot projection, $P \overset{\text{RI, strong RIII}}{\sim} \bigcirc$.

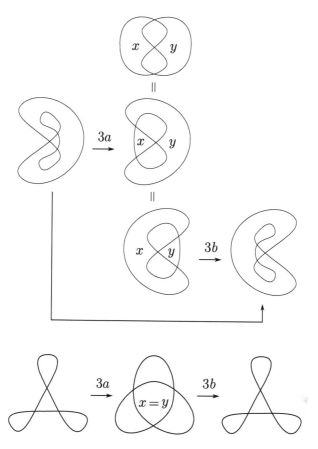

FIGURE 5.14 Case 4-(i), where $m = 3$

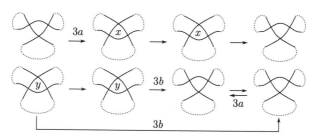

FIGURE 5.15 Case 4-(ii)

By using Theorem 5.2, we can obtain Theorem 5.3. To introduce the statement of Theorem 5.3, we first need to define a *connected sum* of two knot projections P_1 and P_2.

Definition 5.4 Recall the definition of arcs of knot projections (cf. Notation 4.1). Let P be a knot projection. When we go along P, the path starting from a double point to the next encountered double point is called an *arc*. Then, $S^2 \setminus P$ consists of a finite number of disks and one of them is called a *region*. Let P_1 and P_2 be two knot projections. Here, suppose that the ambient spheres corresponding to P_1 and P_2 are oriented. Select an arc from P_i $(i = 1, 2)$, which is denoted by a_i. Let $R_i(P_i)$ be a region in $S^2 \setminus P_i$ where the boundary of R_i contains a_i. For P_2, we choose an infinity point on R_2 and the knot projection is considered to be the knot projection in the disk. The knot projection with the oriented disk is denoted by $D(P_2)$. Then, we place $D(P_2)$ on R_1 where the orientation of the disk $D(P_2)$ and the ambient sphere of P_1 coincides. For arc a_i, take two distinct points $p_1^{(i)}$ and $p_2^{(i)}$ on a_i. Then, there exists a sufficiently small disk such as that shown in the left figure of Figure 5.16 and we replace it with the disk shown in the right figure of Figure 5.16. After this process, we have a new knot projection that is

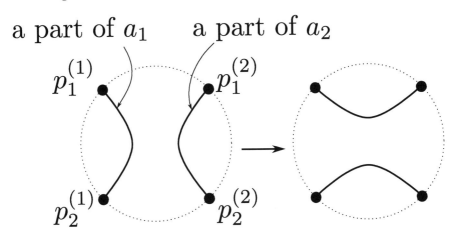

FIGURE 5.16 Operation of exchanging two disks

called a *connected sum* of P_1 and P_2 depending on (a_i, p_i, R_i) $(i = 1, 2)$ and is denoted by $P_1 \natural_{(a,p,R)} P_2$, where $a = (a_1, a_2)$, $p = (p_1^{(1)}, p_2^{(1)}, p_1^{(2)}, p_2^{(2)})$, and $R = (R_1, R_2)$. For simplicity, we often write this as $P_1 \natural P_2$ without specifying which a, p, and R are selected if not required. Note that the definition of $P_1 \natural_{(a,p,R)} P_2$ does not depend on the role of P_1 and P_2 but depends on (a, p, R) as shown above.

Let n be a nonnegative integer. Let S be a set of a finite number of knot projections and is denoted by S. Let $T = \{P_j \mid P_j = \text{an element of } S, (j = 1, 2, \ldots, n)\}$. A connected sum of P_j and P_{j+1} is denoted by

$P_j \natural_{(a(j+1),p(j+1),R(j+1))} P_{j+1}$, where $a = a(j+1)$, $p = p(j+1)$, and $R = R(j+1)$ for two knot projections P_j and P_{j+1}. We consider a connected sum of an ordered knot projection consisting of P_j ($1 \leq j \leq n$) and a sequence of $(a(j), p(j), R(j))$ ($1 \leq j \leq n$); that is,

$$(\ldots(((P_1 \natural_{a(2),p(2),R(2)} P_2) \natural_{a(3),p(3),R(3)} P_3) \natural_{a(4),p(4),R(4)} P_4) \cdots \natural_{a(n),p(n),R(n)} P_n).$$

Such a connected sum is called *a connected sum of knot projections*, each of which is an element of the set S. For any S and T, if we freely consider a sequence $(a(j), p(j), R(j))$ and there is no need to specify the set, then a connected sum of knot projections, each of which is an element of S, is denoted by

$$P_1 \natural P_2 \natural \ldots \natural P_n.$$

Theorem 5.3 ([3]) *Let P be a knot projection and $\bigcirc$ be a trivial knot projection. The following two conditions, (1) and (2), are equivalent:*
(1) P and $\bigcirc$ are related by a finite sequence generated by RI and strong RIII.
(2) P is a connected sum of knot projections, each of which is an element of the set consisting of a trivial knot projection, a knot projection that appears as ∞, and a trefoil knot projection, as shown in Figure 5.17.

FIGURE 5.17 Set consisting of the three knot projections; from left to right: trivial knot projection, knot projection that appears as ∞, and trefoil knot projection

The set consisting of a trivial knot projection, a knot projection that appears as ∞, and a trefoil knot projection is denoted by S_1.

To prove Theorem 5.3, we need uniqueness of the decomposition with respect to each connected sum as follows (Definition 5.5 and Fact 5.4).

Definition 5.5 *If a knot projection cannot be represented as $P_1 \natural P_2$ and is not a trivial knot projection, then the knot projection is called a* prime knot projection.

Fact 5.4 *Let P be a knot projection. Suppose that for P, there exist two sequences of ordered prime knot projections $\{P_i\}_{1 \leq i \leq m}$ and $\{Q_j\}_{1 \leq j \leq n}$ and*

$$P = (\ldots(P_1 \natural_{(a(2),p(2),R(2))} P_2) \natural_{(a(3),p(3),R(3))} P_3) \cdots \natural_{(a(m),p(m),R(m))} P_m),$$

$$Q = (\ldots(Q_1 \natural_{(a'(2),p'(2),R'(2))} Q_2) \natural_{(a'(2),p'(2),R'(2))} Q_3) \cdots \natural_{(a'(n),p'(n),R'(n))} Q_n).$$

Then, by replacing the order of indices $\{j\}$ *(if necessary),*

$$m = n, P_i = Q_i, (a(i), p(i), R(i)) = (a'(i), p'(i), R'(i))(1 \leq i \leq m)$$

.

Now we prove Theorem 5.3.

Proof 5.6 Proof of (1) $\Rightarrow$ (2).

By Theorem 5.2, if we have (1), then there exists a sequence generated by
$(1a)$ and $(s3a)$ from $\bigcirc$ to P.

We prove the claim by an induction with respect to the number of applications n (≥ 0) of Reidemeister moves $(1a)$ or $(3a)$ to $\bigcirc$.

[1] For the case when $n = 0$, $P = \bigcirc$; therefore, we have the claim.

[2] Suppose that the claim holds for the case when $n = k$. Under this assump-
tion, we show the claim for $n = k + 1$. Depending on whether the $(k + 1)$th
Reidemeister move is $(1a)$ or $(3a)$, we consider the following two cases (see
Figure 5.18).

In this proof, for simplicity, a knot projection that appears as ∞ (resp. a
trefoil knot projection) is denoted by ∞ (resp. 3_1).

Case $(1a)$. Refer to Figure 5.18 (1st and 2nd lines). The application of a
single $(1a)$ to a knot projection P is replaced by generating a connected sum
$P\sharp\infty$. With the assumption of the induction and Fact 5.4, P is a connected
sum of knot projections, each of which is an element of S_1. Then, $P\sharp\infty$ is
a connected sum of knot projections, each of which is an element of S_1, as
shown in Figure 5.18.

Case $(s3a)$. Refer to Figure 5.18 (third and fourth lines). For the application of a single $(3a)$, there exist three knot projections P_1', P_2', and P_3' and it
can be written by the following replacement (using Fact 5.4):

$$n = k : P_1'\sharp P_2'\sharp P_3'\sharp\infty\sharp\infty\sharp\infty,$$

$$n = k + 1 : P_1'\sharp P_2'\sharp P_3'\sharp 3_1.$$

Here, there is no intersection between any two of three knot projections P_1',
P_2', and P_3', the reason for which is obtained by Fact 5.5 (Fact 5.5 will be
shown in Chapter 6).

Fact 5.5 (Proposition 6.4) *Let P be a knot projection and $\bigcirc$ be a trivial
knot projection. If P and $\bigcirc$ are related by a finite sequence generated by
RI and strong RIII, then any strong RIII in the sequence satisfies the following
condition.*

*Case 1. If a single strong RIII is written as the left figure on the upper line
of Figure 5.19, then there is no double point between any two parts of (A),
(B), and (C).*

*Case 2. If a single strong RIII is written as the right figure on the upper
line of Figure 5.19, then there is no double point between any two parts of
(A'), (B'), and (C').*

This completes the induction, which shows the claim.

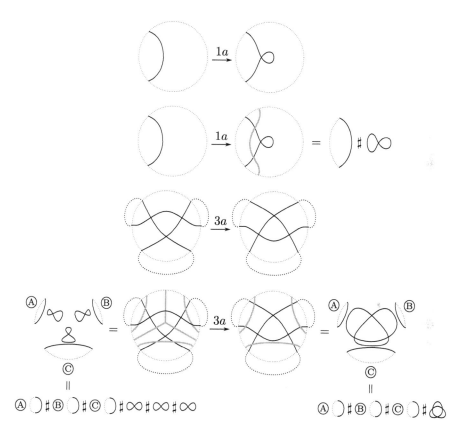

FIGURE 5.18 (1a) (1st line) and (3a) (3rd line). Prime decompositions of two knot projection between before and after (1a) (resp. (3a)) on the 2nd line (resp. 4th line)

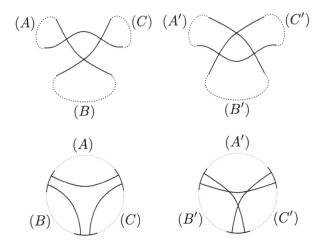

FIGURE 5.19 Parts (A), (B), and (C) and (A'), (B'), and (C')

5.5 OPEN PROBLEM AND EXERCISES

1. (open) Is the map $[f]$ injective? (see the proof of Theorem 5.1; this question was independently given by Seiichi Kamada and Yasutaka Nakanishi).

2. Consider the positive resolution with respect to strong RⅢ and understand that it corresponds to a Δ-unknotting operation (Murakami-Nakanishi [4]).

3. (known result) Write the proof of Fact 5.4.

Bibliography

[1] T. Hagge and J. Yazinski, On necessity of Reidemeister move 2 for simplifying immersed planar curves, *Knots in Poland III. Part III*, 101–110, Banach Center Publ., 103, *Polish Acad. Sci. Inst. Math., Warsaw, 2014.*

[2] N. Ito and Y. Takimura, (1, 2) and strong (1, 3) homotopies on knot projections, *J. Knot Theory Ramifications* **22** (2013), 1350085, 14pp.

[3] N. Ito, Y. Takimura, and K. Taniyama, Strong and weak (1, 3) homotopies on knot projections, *Osaka J. Math.* **52** (2015), 617–646.

[4] H. Murakami and Y. Nakanishi, On a certain move generating link-homology, *Math. Ann.* **284** (1989), 75–89.

[5] O.-P. Östlund, Invariants of knot diagrams and diagrammatic knot invariants, Thesis (Ph.D.)–Uppsala Universitet (Sweden). *ProQuest LLC, Ann Arbor, MI,* 2001. 67pp.

[6] O. Viro, Generic immersions of the circle to surfaces and the complex topology of real algebraic curves. *Topology of real algebraic varieties and related topics*, 231–252, Amer. Math. Soc. Transl. Ser. 2, 173, *Amer. Math. Soc., Providence, RI,* 1996.

CHAPTER 6

Techniques for counting sub-chord diagrams (2015–Future)

CONTENTS

6.1   Chord diagrams 73
6.2   Invariants by counting sub-chord diagrams 75
    6.2.1   Invariant X 75
    6.2.2   Invariant H 76
    6.2.3   Invariant λ 80
6.3   Applications of invariants 82
6.4   Based arrow diagrams 84
6.5   Trivializing number 86
6.6   Exercises 92

T echniques are required to address the classification problems of knot projections. One methods has been suggested in the works of Ito–Takimura [4] (2015). This chapter provides some examples and shows the effectiveness of this technique.

6.1   CHORD DIAGRAMS

A *Gauss word* w of length $2n$ is a two-to-one map from $\hat{2n} = \{1, 2, \ldots, 2n\} \to \mathbb{N}$; that is, each $w^{-1}(i)$ consists of two numbers. Traditionally, a Gauss word has been represented as $w(1)w(2) \cdots \cdot w(2n)$, where each $w(i) \in \hat{n}$. Then we call each $w(i)$ a *letter*. We define cyc and rev by $\mathrm{cyc}(p) \equiv p + 1 \pmod{2n}$ and $\mathrm{rev}(p) \equiv -p + 1 \pmod{2n}$. Two Gauss words, v and w, of length $2n$ are *isomorphic* if there exists a bijection $f : v(\hat{2n}) \to w(\hat{2n})$ satisfying that there exists $t \in \mathbb{Z}$ such that $w \circ (\mathrm{cyc})^t \circ (\mathrm{rev})^\epsilon = f \circ v$ ($\epsilon = 0, 1$). The isomorphism class containing a Gauss word w is denoted by $[w]$. We call every configuration of n pair(s) of points on a circle a *chord diagram*. A chord diagram is presented as a circle with paired points. Usually, two points of each pair on the circle of a chord diagram are connected by a curve that has no self-intersections, called a *chord*.

Note that the isomorphism classes of the Gauss words of length $2n$ have one-to-one correspondence with the chord diagrams, each of which has n chords. In the rest of this book, we identify four expressions shown in Figure 6.1 and we will freely use either one of them depending on situations.

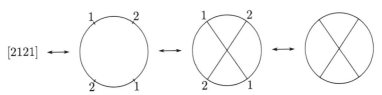

FIGURE 6.1 Four expressions

Let P be a knot projection, that is, a generic immersion $g : S^1 \to S^2$ such that $g(S^1) = P$ and let k be the number of the double points of P. Then we assign a chord diagram to P in the following. First, we take a base point that is not a double point and choose an orientation of P. Starting from the base point, we go along the knot projection once and return back to the base point. When we first encounter a double point d_1, we assign "1" to d_1. Next, we encounter the second double point d_2 and if the double point d_2 has not yet been labeled as "1", we assign "2" to d_2 and if d_2 has label "1," we just pass through d_2. Suppose that $1, 2, \ldots, p$ have been assigned. Then we assign $p+1$ to the next double point that we encounter provided it has not been assigned a positive integer yet. Following the same procedure, we finally assign $1, 2, \ldots, k$ to the double points of P. Here, note that g^{-1}(double point assigned i) consists of two points on S^1 and we assign i to the two points. Then we have a chord diagram represented by g^{-1}(double point assigned 1), g^{-1}(double point assigned 2), $\ldots$, g^{-1}(double point assigned k) on S^1. The chord diagram is denoted by CD_P and is called a *chord diagram of the knot projection P*.

We should be careful that the representation of a chord diagram by chords and a circle is not unique. For instance, a chord diagram of a trefoil knot projection has at least two representations (Figure 6.2).

Definition 6.1 Let CD be a chord diagram. We fix a Gauss word G representing CD. Without loss of generality, we may suppose the set of letters as $G = \{1, 2, \ldots, n\}$. Then, for G, we consider the set of letters as $\{G' \mid G'$ is obtained from G by ignoring some of the letters$\}$. This set is denoted by $\mathrm{Sub}(G)$. An isomorphism class of an element of $\mathrm{Sub}(G)$ is called a *sub-chord diagram*. Let x be a chord diagram. Next, let $\mathrm{Sub}(x)(G) = \{G' \in \mathrm{Sub}(G) \mid [G'] = x\}$. The cardinality of $\mathrm{Sub}(x)(G)$ is denoted by $x(G)$. Let H be another Gauss word representing CD. It is elementary to show that the definition of the isomorphism of the Gauss words implies $x(H) = x(G)$. Thus, we shall

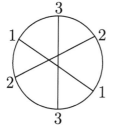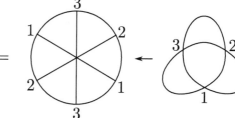

FIGURE 6.2 Two chord diagrams of a trefoil knot projection

denote this number by $x(CD)$. Let P be a knot projection. If $CD = CD_P$, we may denote $x(CD_P)$ simply by $x(P)$. We may also denote $\text{Sub}(x)(G)$ by $\text{Sub}(x)(P)$.

Let $\mathcal{C}$ be the set of the knot projections. Naturally, we consider x as a map $\mathcal{C} \to \mathbb{Z}_{\geq 0}$ such that $P \mapsto x(P)$. By using this definition, $\bigotimes(P)$, $\bigotimes(P)$, and $\bigoplus(P)$ are defined.

6.2   INVARIANTS BY COUNTING SUB-CHORD DIAGRAMS

Let $\mathcal{C}$ be the set of knot projections as above. Let $\mathcal{S}$ be a set where we can easily detect any two different elements s_1 and s_2 as $s_1 \neq s_2$ by using the equality in set $\mathcal{S}$. In this book, an invariant I is a map $I : \mathcal{C} \to \mathcal{S}$. The image $I(\mathcal{C})$ is also called an *invariant*.

6.2.1   Invariant X

Let $X : P \mapsto \mathbb{Z}_{\geq 0}$ be a map defined by

$$X(P) = \begin{cases} 0 & \bigotimes(P) = 0 \\ 1 & \bigotimes(P) \neq 0. \end{cases} \tag{6.1}$$

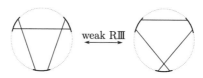

FIGURE 6.3   Difference before and after the application of a single weak RIII

Proposition 6.1 ([4]) *X is invariant under RI and weak RIII.*

Proof 6.1 Seeing the difference between two chord diagrams under RI, any single RI does not increase or decrease the cardinality of $\mathrm{Sub}(\bigotimes)(P)$ (Figure 6.4). In Figure 6.3, no difference can be observed between before and after

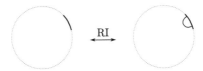

FIGURE 6.4 Difference before and after the application of a single RI

the application of a single weak RIII; $X(P) = 1$ and is fixed under weak RIII. Thus, $X(P)$ is invariant under RI and weak RIII.

6.2.2 Invariant H

Let $H : \mathcal{C} \to \mathbb{Z}_{\geq 0}$ be a map defined by

$$H(P) = \begin{cases} 0 & \bigoplus (P) = 0 \\ 1 & \bigoplus (P) \neq 0. \end{cases} \tag{6.2}$$

Proposition 6.2 ([5]) *A map H is invariant under RI and strong RIII.*

FIGURE 6.5 Difference before and after the application of a single strong RIII

Proof 6.2 First, observe the difference between before and after the application of a single RI (Figure 6.4). Thus, by definition, H is invariant under RI. Second, observe the difference between CD_{P_1} and CD_{P_2} under a single strong RIII (Figure 6.5), where $\bigotimes (P_1) > \bigotimes (P_2)$. Let us compare $\mathrm{Sub}(\bigoplus)(P_1)$ and $\mathrm{Sub}(\bigoplus)(P_2)$. Consider $h \in \mathrm{Sub}(\bigoplus)(P_1) \setminus \mathrm{Sub}(\bigoplus)(P_2)$. Here, h should have an additional chord, denoted by c_h, as shown in Figure 6.6.

FIGURE 6.6 Difference before and after the application of a single strong RⅢ and an additional chord

If there exists a chord c_h, then h containing c_h appears for two elements in Sub($\bigoplus$)(P_1) for every c_h.

For the same c_h, there exists exactly one $h' \in \text{Sub}(P_2) \setminus \text{Sub}(P_1)$ containing c_h.

Hence, we have two cases.

- Case 1: such that there exists no c_h. Sub($\bigoplus$)(P_1) = Sub($\bigoplus$)(P_2). Thus, $H(P_1) = H(P_2)$ for this case.

- Case 2: such that there exists at least one c_h. In this case, we always have $H(P_1) = H(P_2)\ (= 1)$.

This completes the proof.

By using H, we have Proposition 6.3.

Proposition 6.3 ([5]) *Let P be a knot projection. If P and $\bigcirc$ can be related by a finite sequence generated by RI and strong RⅢ, then P cannot have x and y $(1 \leq m \leq 2)$, as shown in Figure 6.7.*

Proof 6.3 For a knot projections P_1 and P_2, as shown in Figure 6.7, $H(P_1) = H(P_2) = 1 \neq 0 = H(\bigcirc)$, as shown in Figure 6.8.

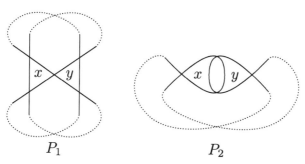

$$P_1 \qquad\qquad P_2$$

FIGURE 6.7 P_1 and P_2. Dotted arcs indicate fragment connections

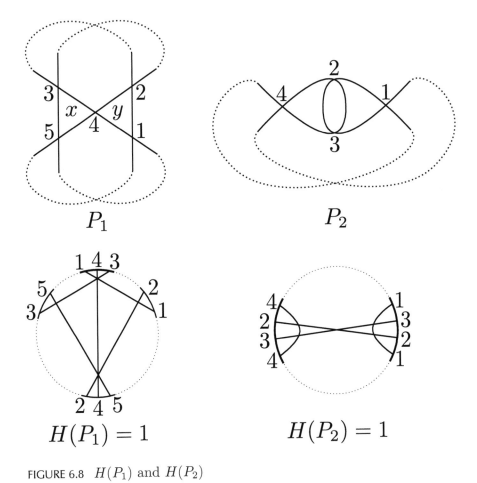

FIGURE 6.8 $H(P_1)$ and $H(P_2)$

We also have Proposition 6.4, which is Fact 5.5. Recall the definition of *strong* 3-gons.

Definition 6.2 Let P be a knot projection with at least one 3-gon. If by selecting an orientation of P, we can orient the 3-gon such that it coincides with the orientation of P, then the 3-gon is called a *strong* 3-gon (see Figure 6.9). For every strong 3-gon, we obtain an expression, as shown in the bottom line of Figure 6.9 (expression Ⅱ) using dotted sub-curves that indicate the connections of the three paths. Such three dotted sub-curves are called *underlying sub-curves* of a strong 3-gon.

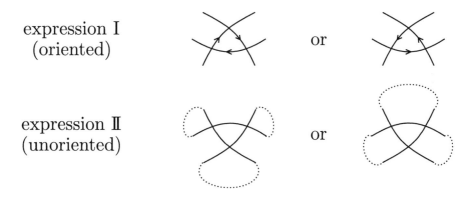

expression I (oriented)   or

expression Ⅱ (unoriented)   or

FIGURE 6.9 Strong 3-gons

Proposition 6.4 *Using the above terminology, we have the following. Let P be a knot projection and $\bigcirc$ be a trivial knot projection. Suppose that P and $\bigcirc$ are related by a finite sequence generated by RI and strong RⅢ. Then, if P has a 3-gon, then there is no intersection between any two underlying sub-curves.*

Proof 6.4 See Figure 6.10. Without loss of generality, it is sufficient to show the claim for (A) and (B) of Case 1 and (A′) and (B′) of Case 2.

Let Q be a knot projection having a 3-gon for Case 1 or Case 2, expressed as the upper line in Figure 6.10. Then, if there exists a chord connecting (A) and (B) (resp. (A′) and (B′)), then using the bottom line of Figure 6.10, we can see that $H(Q) = 1$.

However, $H(\bigcirc) = 0$. Thus, Q and $\bigcirc$ cannot be related by a finite sequence generated by RI and strong RⅢ. Therefore, if a knot projection P can be related to $\bigcirc$ by a finite sequence generated by RI and strong RⅢ, there will be no intersection between any two underlying sub-curves.

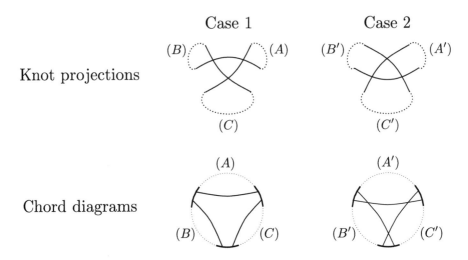

FIGURE 6.10 3-gons and the corresponding chord diagrams

6.2.3 Invariant λ

Definition 6.3 Let I be an invariant. We say that I is *additive* if for any two knot projections P_1 and P_2,

$$I(P_1 \sharp P_2) = I(P_1) + I(P_2).$$

Let λ be a map $\frac{1}{4}\left(3\bigoplus - 3\bigotimes + \bigotimes\right)$.

Proposition 6.5 ([4]) *Let P be a knot projection. Let*

$$\lambda(P) = \frac{1}{4}\left(3\bigoplus(P) - 3\bigotimes(P) + \bigotimes(P)\right).$$

Then $\lambda(P)$ is invariant of P under RI and strong RIII and is additive.

Remark 6.1 [4] shows that $\lambda(P) \in \mathbb{Z}$ for every knot projection P.

Now we show Proposition 6.5.

Proof 6.5 By Figure 6.4, $\bigoplus$, $\bigotimes$, and $\bigotimes$ are invariant under RI. Thus, we consider a single strong RIII for each $x = \bigoplus$, $\bigotimes$, and $\bigotimes$.

- $\bigoplus$. For a single strong RIII between P_1 and P_2, set P_1 and P_2 such that $\bigotimes(P_1) > \bigotimes(P_2)$ holds. Denote by c_h a chord that appears in Figure 6.6, which does not change under strong RIII. For every c_h, the difference of $\bigoplus (P)$ is exactly $2 - 1 = 1$ (see the proof of Proposition 6.2). If k chords of type c_h are present in a chord diagram of P_1, the difference is exactly k. The reason is described as follows. A single strong RⅢ changes three chords. The three chords are called RⅢ-chords and labeled as $r_i(P)$ ($i = 1, 2, 3$ and $P = P_1, P_2$) here. Now we can obtain a one-to-one correspondence between $r_i(P_1)$ and $r_i(P_2)$ by sliding the RⅢ-chords such that $r_i(P)$ is maintained and the position among non-RⅢ-chords is not changed.

Thus, $h \in \mathrm{Sub}(\bigoplus)(P_1)$ has two chords of type c_h if and only if $h \in \mathrm{Sub}(\bigoplus)(P_2)$ has the two chords of c_h. Thus, if $h \in \mathrm{Sub}(\bigoplus)(P_1)$ but $h \notin \mathrm{Sub}(\bigoplus)(P_2)$ and h contains c_h, the number of chords of type c_h is exactly equal to one. Then, in this case, there is a one-to-one correspondence between h and c_h. Thus, if there are k chords of type c_h, $\bigoplus(P_1) - \bigoplus(P_2) = k$.

- $\bigotimes^{*}$. Set P_1, P_2, c_h, $r_i(P)$, and the identification of $r_i(P_1)$ and $r_i(P_2)$ by the same definition as mentioned above for $\bigoplus$. If there is no chord of type c_h, $\bigotimes^{*}(P_1) - \bigoplus(P_2) = 1$. If there exist k chords of type c_h, we check the difference. If $\bigotimes^{*} \in \mathrm{Sub}(\bigotimes^{*})(P_1)$ (resp. $\mathrm{Sub}(\bigotimes^{*})(P_2)$) consists of two c_h and $r_i(P_1)$ (resp. $r_i(P_2)$), this $\bigotimes^{*}$ satisfies $\bigotimes^{*} \in \mathrm{Sub}(\bigotimes^{*})(P_2)$ (resp. $\mathrm{Sub}(\bigotimes^{*})(P_1)$). If $\bigotimes^{*} \in \mathrm{Sub}(\bigotimes^{*})(P_1)$ consists of one c_h and two $r_i(P_1)$, $\bigotimes^{*} \notin \mathrm{Sub}(\bigotimes^{*})(P_2)$. Thus, there is one-to-one correspondence between $\bigotimes^{*} \in \mathrm{Sub}(\bigotimes^{*})(P_1) \setminus \mathrm{Sub}(\bigotimes^{*})(P_2)$ and c_h except for the cases when $\bigotimes^{*}$ consists of three chords $r_i(P_1)$ ($i = 1, 2, 3$) and $\bigotimes^{*}$ consists of two RⅢ-chords and a single chord of type c_h. Thus, there are k chords of type c_h, $\bigotimes^{*}(P_1) - \bigotimes^{*}(P_2) = k + 1$.

- $\bigotimes$. Set P_1, P_2, and $r_i(P)$ as mentioned above. It is easy to see $\bigotimes$ such that $\bigotimes \in \mathrm{Sub}(\bigotimes)(P_1) \setminus \mathrm{Sub}(\bigotimes)(P_2)$ is $\bigotimes$ consisting of a pair $(r_i(P_1), r_j(P_2))$ ($i \neq j$). Thus, $\bigotimes(P_1) - \bigotimes(P_2) = +3$.

In summary, $4\lambda(P_1) - 4\lambda(P_2) = 3k - 3(k + 1) + 3 = 0$, which implies the invariance of $\lambda(P)$.

Finally, we show the additivity property. For each $x = \bigotimes, \bigoplus,$ and $\bigotimes^{*}$, we have $x(P \sharp P') = x(P) + x(P')$. Thus, we have $\lambda(P \sharp P') = \lambda(P) + \lambda(P')$.

6.3 APPLICATIONS OF INVARIANTS

Recall that a single RIII can be divided into two kinds of local moves: strong and weak. We know the following two results. Denote by $P \overset{x,y}{\sim} P'$ or $P \overset{x}{\sim} P'$ that two knot projections P and P' can be related by a finite sequence generated by two types of Reidemeister moves x and y. Let 1b be a single RI decreasing the number of double points. We also recall the definition of a *connected sum* (for the definition, see Chapter 5).

Theorem 6.1 ([3]) *Let* $\bigcirc$ *be a trivial knot projection and* P *be a knot projection.* $P \overset{RI,\ weak\ RIII}{\sim} \bigcirc \Leftrightarrow P \overset{1b}{\sim} \bigcirc.$

Theorem 6.2 ([5]) *Let* P *be a knot projection.* $P \overset{RI,\ strong\ RIII}{\sim} \bigcirc \Leftrightarrow P$ *is a connected sum of knot projections, each of which is an element of the set consisting of a trivial knot projection, a knot projection that appears as* ∞, *and a trefoil knot projection, as shown in Figure 6.11.*

FIGURE 6.11 Set consisting of three elements

Now, we reprove these results.
(Proof of Theorem 6.1).

Proof 6.6 ($\Leftarrow$) If $P \overset{RI}{\sim} \bigcirc$, then $P \overset{RI,\ RIII}{\sim} \bigcirc$. ($\Rightarrow$). If $P \overset{RI}{\sim} \bigcirc$, $X(P) = X(\bigcirc) = 0$. Then, there exists a finite sequence generated by RI and RIII while $X(\cdot) = 0$. However, this sequence must not contain RIII. If the sequence contains a single RIII, the image of X will be equal to 1, which contradicts the condition $X(\cdot) = 0$. Thus, we have a finite sequence generated by RI, which implies $P \overset{RI}{\sim} \bigcirc$.

If $P \overset{RI}{\sim} \bigcirc$, by Khovanov's theorem in Chapter 4, we have $P \overset{1b}{\sim} \bigcirc$. This completes the proof.

Here, we introduce another way to directly find a finite sequence from P to $\bigcirc$ generated by RI, where this sequence has decreasing the number of double points when $X(P) = 0$ (see Figure 6.12). If a chord and a part of the chord diagram of S^1 form a biangle that does not contain any chords, this chord is called an *innermost chord*. As the first step, when $X(P) = 0$, because there is a finite number of chords, we can find an innermost chord in CD_P. This corresponds to a 1-gon in a knot projection. Then, we apply a single RI decreasing the number of double points to this 1-gon. We repeatedly perform this process until there exists no innermost chord, which implies $\bigcirc$.

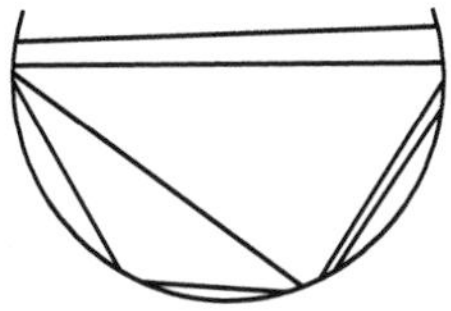

FIGURE 6.12   Chords without intersections

(Proof of Theorem 6.2). (⇐) In this direction, an operation of a connected sum corresponds to a sequence generated by RI and strong RIII (see
Figure 6.13). For the first line in Figure 6.13, the operation on a curve that

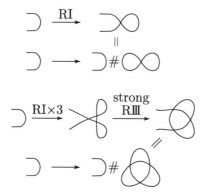

FIGURE 6.13   Connected sum and a sequence of Reidemeister moves

appears as ∞ can be replaced by a single RI. For the second line, the operation
on a trefoil projection can be replaced by three RI and a single strong RIII.

   If a knot projection satisfies the condition on the right-hand side of the
claim, we construct a finite sequence generated by RI and strong RIII from P
to $\bigcirc$.

(⇒). Suppose that $P \overset{\text{RI, strong RIII}}{\sim} \bigcirc$. Then, $H(P) = H(\bigcirc) = 0$ and $\lambda(P) = \lambda(\bigcirc) = 0$. Thus, $\boxplus(P) = 0$ and $\bigotimes(P) = \bigtimes(P)$.

   The case when $\bigotimes(P) = 0$ has already been considered above (Theorem 6.1), $P \overset{1b}{\sim} \bigcirc$.

   If we find $\bigotimes$, then we add a chord to it. For $\boxplus(P) = 0$, this chord
added to $\bigotimes$ becomes $\bigotimes$ or it does not intersect with $\bigotimes$. For every $\bigotimes$,
this condition should be satisfied. Then, P should be the connected sum of
the set $\{T_i(i \in \mathbb{Z}_{\geq 0})\}$, where T_i is defined by Figure 6.14.

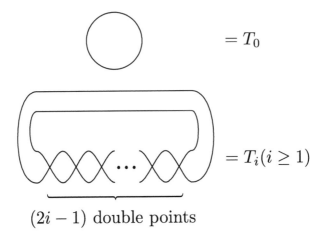

$$(2i - 1) \text{ double points}$$

FIGURE 6.14 T_i: $(2, i)$-torus projection.

Here, recall the property (additivity) of λ, i.e., $\lambda(P \sharp P') = \lambda(P) + \lambda(P')$. For any $i \geq 3$, $4\lambda(T_i)$ is

$$-3\binom{2i-1}{3} + \binom{2i-1}{2}$$
$$= -(2i-1)(i-1)(2i-3) + (2i-1)(i-1) \tag{6.3}$$
$$= (2i-1)(i-1)(-2i+4) < 0.$$

On the other hand, $\lambda(T_i) = 0$ for $i = 0, 1, 2$. Thus, if P satisfies that $H(P) = \lambda(P) = 0$, then P is a connected sum of knot projections, each of which is an element of set $\{T_0, T_1, T_2\}$. This completes the proof.

6.4 BASED ARROW DIAGRAMS

Using chord diagrams, we prove Proposition 5.1. Before proving this proposition, we define the notion of *based arrow diagrams*.

Definition 6.4 (based arrow diagrams) Let v be a Gauss word of length $2n$ such that $v = v(1)v(2) \cdots v(2n)$. For each letter k of v, there exists i and j such that $v(i) = v(j) = k$. Then, we distinguish the two letters $v(i)(= k)$ and $v(j)(= k)$ by calling one k a *tail* and the other a *head*. We express the assignments by adding extra information to v, i.e., we add "→" to every tail. Then the new word obtained from v is denoted by v^* and is called an *oriented Gauss word*. We call every letter of an oriented Gauss word *oriented letter*. Without loss of generality, we may suppose $v(\hat{2n}) = \{1, 2, \ldots, n\}$. Then, clearly, $v^*(\hat{2n}) = \{1, 2, \ldots n, \bar{1}, \bar{2}, \ldots, \bar{n}\}$. Let w^* be another oriented Gauss word of length $2n$ obtained from w. Two oriented Gauss words are *isomorphic* if there exists a bijection $f : v(\hat{2n}) \to w(\hat{2n})$ such that $w^* \circ (\text{rev})^\epsilon = f \circ v^*$

$(\epsilon = 0, 1)$ where $f^* : v^*(\widehat{2n}) \to w^*(\widehat{2n})$ is the bijection such that $f^*(i) = f(i)$ and $f^*(\bar{i}) = \overline{f(i)}$ $(i = 1, 2, \ldots, n)$, and where rev $: \widehat{2n} \to \widehat{2n}$ is as in Section 6.1. The isomorphisms obtain a equivalence relation on the oriented Gauss words. The isomorphism class containing an oriented Gauss word v^* is denoted by $[[v^*]]$. Recall that a chord diagram is a configuration of n pair(s) of points on a circle. For a chord diagram, a point that does not coincide with a point of n pair(s) is placed on the circle of the chord diagram and is called a *base point*. Further, for each pair of the chord diagram, we distinguish the two points such that one is a starting point and the other is an end point. Usually, two points of each pair are connected by an oriented chord, called an *arrow*. A *based arrow diagram* is a chord diagram with a base point where each pair consists of a starting point and an end point. For a based arrow diagram consisting of arrows $\alpha_1, \alpha_2, \ldots, \alpha_k$, the based arrow diagram consisting of a subset of $\{\alpha_1, \alpha_2, \ldots, \alpha_k\}$ is called a *sub-arrow diagram* of the arrow diagram.

Note that the isomorphism classes of oriented Gauss words have one-to-one correspondence with the based arrow diagrams in such a way that each tail corresponds to a starting point and each head corresponds to an end point (see Figure 6.15).In the rest of this book, we identify these four expressions and will freely use them depending on situations.

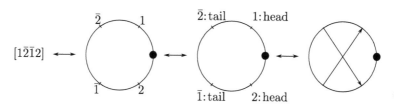

FIGURE 6.15 Four expressions

A crossing of a knot diagram has over-/underinformation. For a crossing point, we obtain the orientation from an overpath to an under path at the crossing. A knot diagram with a base point that is not a crossing is called a *pointed knot diagram*. As we know the correspondence between a knot projection and a chord diagram, we can understand that a pointed knot diagram can be used to determine a based arrow diagram.

Now, we consider one more structure. Let $\tilde{K}$ be a pointed knot diagram. The based arrow diagram corresponding to $\tilde{K}$ is denoted by $AD_{\tilde{K}}$. For a pointed knot diagram, we select an orientation of the knot diagram. Every crossing has an over-/underinformation and consists of a transverse double point. For a crossing, the tangent vector of the overpath (resp. under path) is denoted by t_o (resp. t_u). For (t_o, t_u), we consider the rotation from t_o to t_u (in the tangent plane on a sphere) and this rotation determines the orientation. If (t_o, t_u) is oriented counterclockwise, the crossing is a *positive crossing*; otherwise, the crossing is a *negative crossing*. Using this information, we assign ± 1

to every arrow. Thus, for $\tilde{K}$, every sub-based arrow diagram has a sign that is the product of all signs of the arrows. Let x be a based arrow diagram and let A be a sub-based arrow diagram that is isomorphic to x in $AD_{\tilde{K}}$. The sign of A is denoted by sgn $A(x, \tilde{K})$. Here, we review a well-known result obtained by Polyak–Viro in knot theory.

Theorem 6.3 ([6]) *Let K be a knot and $\tilde{K}$ be a pointed knot diagram where a base point is chosen arbitrarily.*

$$\sum_A \text{sgn } A(x, \tilde{K})$$

does not depend on the choice of $\tilde{K}$ (i.e., on the knot diagram and the base point) and is a knot invariant. Thus, $\sum_A$ sgn $A(x, \tilde{K})$ is denoted by ⊗ (K).

Here, using Theorem 6.3, we obtain the proof of Fact 5.1.

Proof 6.7 Every crossing of a positive knot K is a positive crossing, ⊗$(K) \geq 0$ by definition. When we omit the base point and the orientation of the chords, there exists a sub-arrow diagram of type ⊗ if there is type ⊗ . This is because ⊗ (K) does not depend on the choice of the base point of $\tilde{K}$ by Theorem 6.3. Therefore, if there exists at least one sub-chord diagram of type ⊗ embedded into the chord diagram of a knot projection determined by a knot diagram of K, then ⊗ $(K) > 0$.

Therefore, if K is a trivial knot, ⊗(K) should be 0, which implies a knot projection P, determined by a knot diagram of K, which has no sub-chord diagram of type ⊗ .

When we have a chord diagram $X(P) = 0$, as shown in Figure 6.12, we have $P \overset{1b}{\sim}$ ○. In our case, we have chords that have signs and orientations; however, the discussion is the same. Thus, we have the claim.

6.5 TRIVIALIZING NUMBER

Hanaki introduced the trivializing number to consider its applications to DNA knots [1]. In topology, the trivializing number can be used for the study of the unknotting number of knots [2]. Here, [5] showed that Hanaki's trivializing number is invariant under RI and weak RⅢ. Let CD be a chord diagram and CD_P be a chord diagram of a knot projection P. If a chord diagram CD does not have ⊗ as a sub-chord diagram, it is called a *trivial chord diagram*. We define the trivializing number $tr(P)$ of knot projection P by

$$tr(P) = \min\{\text{the number of erased chords until a trivial chord diagram}$$
$$\text{is obtained from } CD_P\}.$$

Remark 6.2 Note that this definition is *not* Hanaki's original definition, but for simplicity, we define this using CD_P.

In [5], Theorem 6.4 is shown:

Theorem 6.4 ([5]) *Let P be a knot projection and let $tr(P)$ be the trivializing number of P.*

1. *$tr(P)$ is invariant under RI and weak RⅢ.*

2. *$tr(P)$ changes by ± 2 or 0 under strong RⅢ.*

Proof 6.8 • (Invariance under RI) Recall that $\bigotimes(P)$ is invariant under RI. This implies that $tr(P)$ is invariant under RI.
• (Invariance under weak RⅢ) Let P_1 and P_2 be knot projections such that $P_1 \overset{\text{weak RⅢ}}{\sim} P_2$ by a single weak RⅢ, as shown in Figure 6.16. In order to show that $tr(P_1) = tr(P_2)$, we will show that (1) $tr(P_1) \geq tr(P_2)$ and (2) $tr(P_1) \leq tr(P_2)$.
(1) Proof of $tr(P_1) \geq tr(P_2)$. Let $c_1, c_2, \ldots, c_{t(P_1)}$ be chords in CD_{P_1} such that if we erase chords $c_1, c_2, \ldots, c_{tr(P_1)}$ from CD_{P_1}, then the resulting chord diagram will be a trivial chord diagram. Here, the three chords shown in Figure 6.16 are denoted by a, b, and c. We show the claim for each of three cases as follows:

1. Case 1: one element of $\{a, b, c\}$ belongs to $\{c_i\}_{i=1}^{tr(P_1)}$. The selected element should be either b or c.

   • Suppose that $b \in \{c_i\}_{i=1}^{tr(P_1)}$. Without loss of generality, set $c_1 = b$. In this case, deleting $\{e\} \cup \{c_i'\}_{i=2}^{tr(P_1)}$ gives a trivial chord diagram in CD_{P_2} where c_i has one-to-one correspondence with c_i' ($2 \leq i \leq tr(P_1)$). By the minimality of $tr(P_2)$, $tr(P_1) \geq tr(P_2)$.

   • Suppose that $c \in \{c_i\}_{i=1}^{tr(P_1)}$. In this case, the proof is obtained in the same way as that obtained for b, as shown above.

2. Case 2: two elements of $\{a, b, c\}$ belong to $\{c_i\}_{i=1}^{tr(P_1)}$.

   • Suppose that $b, c \in \{c_i\}_{i=1}^{tr(P_1)}$. Without loss of generality, let $c_1 = b$ and $c_2 = c$. In this case, deleting $\{d, f\} \cup \{c_i'\}_{i=3}^{tr(P_1)}$ gives a trivial chord diagram, where c_i has one-to-one correspondence with c_i' ($3 \leq i \leq tr(P_1)$). By the minimality of $tr(P_2)$, $tr(P_1) \geq tr(P_2)$.

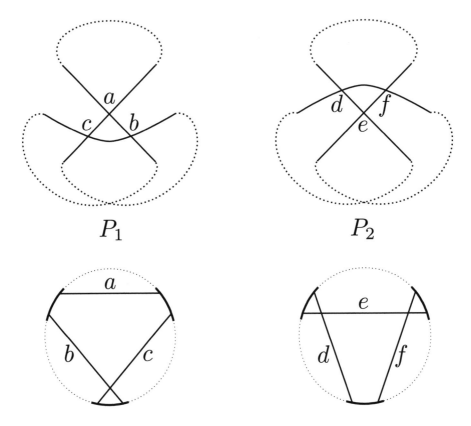

FIGURE 6.16 Two knot projections P_1 and P_2 under single weak RⅢ (upper line) and pair of CD_{P_1} and CD_{P_2} (lower line)

- Suppose that $a, b \in \{c_i\}_{i=1}^{tr(P_1)}$. Without loss of generality, let $c_1 = a$ and $c_2 = b$. In this case, by considering $\{e, d\} \cup \{c_i'\}_{i=3}^{tr(P_1)}$ as shown above, we can have $tr(P_1) \geq tr(P_2)$.

- Suppose that $c, a \in \{c_i\}_{i=1}^{tr(P_1)}$. Without loss of generality, let $c_1 = c$ and $c_2 = a$. In this case, by considering $\{e, f\} \cup \{c_i'\}_{i=3}^{tr(P_1)}$ as shown above, we can have $tr(P_1) \geq tr(P_2)$.

3. Case 3: three elements of $\{a, b, c\}$ belong to $\{c_i\}_{i=1}^{tr(P_1)}$. Without loss of generality, let $c_1 = a$, $c_2 = b$, and $c_3 = c$. By deleting $\{d, e, f\} \cup \{c_i'\}_{i=4}^{tr(P_1)}$, we obtain a trivial chord diagram, where c_i has one-to-one correspondence with c_i' ($4 \leq i \leq tr(P_1)$). The minimality of $tr(P_2)$ implies that $tr(P_1) \geq tr(P_2)$.

Similarly, we can obtain $tr(P_1) \leq tr(P_2)$ and thus, $tr(P_1) = tr(P_2)$.
- (Difference by a single strong RIII)

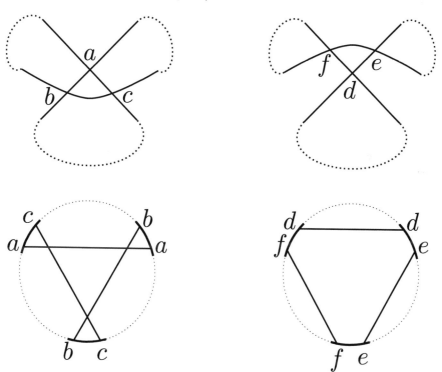

FIGURE 6.17 Two knot projections P_1 and P_2 under single strong RIII (upper line) and pair of CD_{P_1} and CD_{P_2} (lower line)

See Figure 6.17.

- $tr(P_1) - tr(P_2) \geq 2$. Let $\{c_i\}_{i=1}^{tr(P_1)}$ denote the set of chords obtaining the trivializing number of $tr(P_1)$. We consider the following cases.

  1. Case 1: $\{c_i\}_{i=1}^{tr(P_1)}$ contains a, b. Without loss of generality, we can set $c_1 = a$ and $c_2 = b$. In this case, by deleting $\{d, f\} \cup \{c_i'\}_{i=3}^{tr(P_1)}$, we obtain a trivial chord diagram, where c_i' has one-to-one corresponds with c_i $(3 \leq i \leq tr(P_1))$. Then, by the minimality of $tr(P_2)$, $tr(P_1) - 2 \geq tr(P_2)$. For $\{c_i\}_{i=1}^{tr(P_1)}$ that contains the pair of b and c or a and c, we can give $tr(P_1) - tr(P_2) \geq 2$ in the similar way.

  2. Case 2: $\{c_i\}_{i=1}^{tr(P_1)}$ contains a, b, and c. Without loss of generality, we can set $c_1 = a$, $c_2 = b$, and $c_3 = c$. In this case, by deleting $\{d, e, f\} \cup \{c_i\}_{i=4}^{tr(P_1)}$, we obtain a trivial chord diagram where c_i' one-to-one corresponds to c_i $(4 \leq i \leq tr(P_1))$. Then, by the minimality of $tr(P_2)$, $tr(P_1) \geq tr(P_2)$.

In summary, $tr(P_1) - tr(P_2) \geq 0$ or 2.

Next, we prove $tr(P_1) - tr(P_2) \leq 0$ or 2.

  1. $\{c_i'\}_{i=1}^{tr(P_2)}$ does not contain any of d, e, or f. In this case, by deleting $\{a, b\} \cup \{c_i\}_{i=1}^{tr(P_2)}$, we obtain a trivial chord diagram, where c_i' has one-to-one correspondence with c_i $(1 \leq i \leq tr(P_2))$. Then, by the minimality of $tr(P_1)$, $tr(P_1) \leq tr(P_2) + 2$.

  2. $\{c_i'\}_{i=1}^{tr(P_2)}$ contains d. Without loss of generality, we can set $c_1' = d$. In this case, by deleting $\{a, b\} \cup \{c_i\}_{i=2}^{tr(P_2)}$, we obtain a trivial chord diagram, where c_i' has one-to-one correspondence with c_i $(2 \leq i \leq tr(P_2))$. Then, by the minimality of $tr(P_1)$, $tr(P_1) \leq tr(P_2) + 1$. Using the symmetry property, if d is replaced by e or f, we can show the same inequality.

  3. $\{c_i'\}_{i=1}^{tr(P_2)}$ contains d and e. With loss of generality, we can set $c_1' = d$ and $c_2' = e$. In this case, by deleting $\{a, b\} \cup \{c_i\}_{i=3}^{tr(P_2)}$, we obtain a trivial chord diagram, where c_i' has one-to-one correspondence with c_i $(3 \leq i \leq tr(P_2))$. Then, by the minimality of $tr(P_1)$, $tr(P_1) \leq tr(P_2)$. Using the symmetry property, if pair (d, e) is replaced by (e, f) or (f, e), we can show the same inequality.

  4. $\{c_i'\}_{i=1}^{tr(P_2)}$ contains d, e, and f. With loss of generality, we can set $c_1' = d$, $c_2' = e$, and $c_3' = f$. In this case, by deleting $\{a, b, c\} \cup \{c_i\}_{i=4}^{tr(P_2)}$, we obtain a trivial chord diagram, where c_i' has one-to-one correspondence with c_i $(4 \leq i \leq tr(P_2))$. By the minimality of $tr(P_1)$, $tr(P_1) \leq tr(P_2)$.

In summary, $tr(P_1) - tr(P_2) \leq 2$. We already show that $tr(P_1) - tr(P_2) \geq 0, 2$ and then $0 \leq tr(P_1) - tr(P_2) \leq 2$.

Hence, the proof is completed.

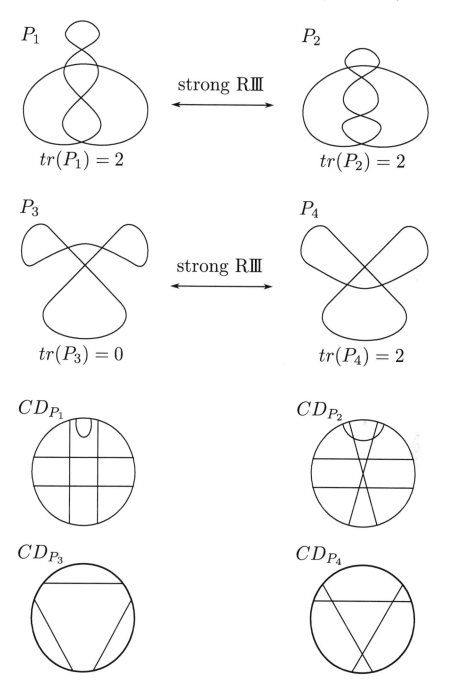

FIGURE 6.18 Examples of strong RⅢ and the trivializing number (upper two lines). Corresponding chord diagrams used to compute trivializing numbers (bottom line)

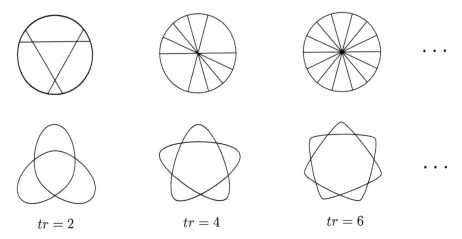

$$tr = 2 \qquad\qquad tr = 4 \qquad\qquad tr = 6$$

FIGURE 6.19 Sequence of knot projections P_m such that $tr(P_m) = 2m$

Example 6.1 Figure 6.18 shows two pairs of knot projections. The pair P_1 and P_2 satisfies $tr(P_1) = tr(P_2) = 2$ (upper line) under a single strong RIII. The other pair P_3 and P_4 satisfies $tr(P_3) = 2$ and $tr(P_3) = 4$ under a single strong RIII (lower line) (Takimura showed these examples to the author).

Example 6.2 A prime knot projection is a non-trivial knot projection that cannot be represented as a connected sum of two non-trivial knot projections. By definition, for every even nonnegative integer $2m$, there exists a prime knot projection P such that $tr(P) = 2m$ (see 6.19; this fact appears in Hanaki's paper [1, Theorem 1.11]).

6.6 EXERCISES

1. Prove Theorem 6.3. Hint: write down Reidemeister moves in terms of the chord diagram.

2. Prove that a single $w2a$ increases the trivializing number by 2. Hint: compare the chord diagrams.

Bibliography

[1] R. Hanaki, Pseudo diagrams, links and spatial graphs, *Osaka J. Math.* **47** (2010), 863–883.

[2] R. Hanaki, Trivializing number of knots, *J. Math. Soc. Japan* **66** (2014), 435–447.

[3] N. Ito and Y. Takimura, (1, 2) and weak (1, 3) homotopies on knot projections, *J. Knot Theory Ramifications* **22** (2013), 1350085, 14pp.

[4] N. Ito and Y. Takimura, Sub-chord diagrams of knot projections, *Houston J. Math.* **41** (2015), 701–725.

[5] N. Ito, Y. Takimura, and K. Taniyama, Strong and weak (1, 3) homotopies on knot projections, *Osaka J. Math.* **52** (2015), 617–646.

[6] M. Polyak and O. Viro, Gauss diagram formulas for Vassiliev invariants, *Internat. Math. Res. Notices* **1994,** 445ff., approx. 8pp. (electronic).

CHAPTER 7

Hagge–Yazinski Theorem (Necessity of RII)

CONTENTS

7.1   Hagge–Yazinski Theorem showing the non-triviality of the equivalence classes of knot projections under RI and RⅢ   96
     7.1.1   Preliminary 96
     7.1.2   Box 97
     7.1.3   Structure of the induction 97
     7.1.4   Moves $1a$, $1b$, and RⅢ inside a rectangle 98
     7.1.5   Moves $1a$, $1b$, and RⅢ outside the rectangles 98
         7.1.5.1 $1a$ 98
         7.1.5.2 $1b$ 98
         7.1.5.3   RⅢ 99
7.2   Arnold invariants 100
7.3   Exercises 104

The first break through, i.e., the first example of a non-trivial homotopy equivalence class under RI and RⅢ, was obtained by Hagge and Yazinski [1]. They first answered Östlund's question, i.e., they showed that there exists a knot projection such that the knot projection cannot be related to a trivial knot projection by a finite sequence generated by RI and RⅢ [3]. Surprisingly, they did not use or obtain any invariant to prove their theorem, which is important to note. Although a new topological invariant under a Reidemeister move is often useful to classify knot projections by using the other Reidemeister moves, it is important for us to find a new invariant. On the other hand, proving the impossibility (i.e., proving that there exists a knot projection that cannot be related to a trivial knot projection) without using an invariant may not be easy to understand.

Hagge and Yazinski understood the characteristic of a special curve and proved the impossibility, which is an invaluable case and worth learning.

7.1 HAGGE–YAZINSKI THEOREM SHOWING THE NON-TRIVIALITY OF THE EQUIVALENCE CLASSES OF KNOT PROJECTIONS UNDER RI AND R$\mathrm{I\!I\!I}$

Hagge and Yazinski proved the non-triviality of equivalence classes of knot projections under RI and R$\mathrm{I\!I\!I}$.

Theorem 7.1 ([1]) *Let P_{HY} be the knot projection, as shown in Figure 7.1. There is no finite sequence generated by RI and R$\mathrm{I\!I\!I}$ between P_{HY} and a trivial knot projection.*

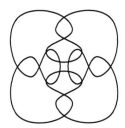

FIGURE 7.1 P_{HY}

In the following, we describe the proof obtained by Hagge and Yazinski.

7.1.1 Preliminary

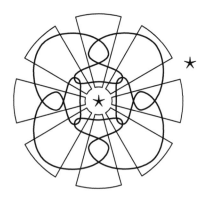

FIGURE 7.2 Eight boxes and $\star$-regions on the knot projection P_{HY}

Consider a part of a knot projection. A connected part with no intersections and its one-component is called a *simple arc*. Knot projection P_{HY} can
be decomposed into eight boxes and simple arcs, as shown in Figure 7.2. A
region marked as $\star$ is called a $\star$-*region*.

7.1.2 Box

In the following, we define the notion of *boxes* for other knot projections.

Let l be a positive integer. Suppose that there exist l sections on a knot
projection. Each section is represented as a rectangle. The sub-curves in the
sections are named as follows. The rectangle consists of four sides; each of the
two-faced sides has three endpoints named by $1, 1, 3$ or $2, 2, 3$. Here, for each
rectangle, we start from label 1 (resp. 2) and reach another label 1 (resp. 2)
on the same side; the sub-curve obtained is labeled as 1 (resp. 2), which is
called strand 1 (resp. 2). If a sub-curve in a rectangle starts from label 3 and
returns to the other side to label 3, then it is called strand 3.

For the rectangles, we consider the following three conditions:

1. (condition 1) In every rectangle, exactly two double points exist between
   strand 1 and strand 2. One of the two endpoints of strand 3 is on the
   left and the other is on the right of a rectangle.

2. (condition 2) There are no double points outside the rectangles. For two
   neighboring rectangles, strand 3 on the right is connected to strand 1 on
   the left of a rectangle and strand 3 on the left is connected to strand 2
   on the right of a rectangle. Because there exist exactly three endpoint
   in each side, strand 1 on the left is connected to strand 2 on the right
   of a rectangle.

3. (condition 3) There exist two disjoint regions satisfying the following:
   the boundary of each region is m-gon $(m \geq 4)$ and each region contains l
   rectangles' sides, each of which has no strand endpoints. The two regions
   are called $\star$-regions.

We call the above three conditions the *Box Rule*. If there exists a finite number
of rectangles satisfying the Box Rule, then the rectangles are called *boxes*.

7.1.3 Structure of the induction

Now, we start the proof of Theorem 7.1 obtained by Hagge–Yazinski.

Proof 7.1 We consider the induction with respect to the number of operations, n, generated by RI and RIII for P_{HY}, i.e., we show Lemma 7.1 for every
n.

Lemma 7.1 *Let P_n be a knot projection obtained from P_{HY} through n operations, each of which is either RI or RIII where $P_0 = P_{HY}$. For every n, there*

exist eight rectangles on P_n such that P_0 is sphere isotopic to P_n outside the rectangles and these eight rectangles satisfy the Box Rule.

In the following, note that the Box Rule does not require the order of 1, 2, and 3 to be fixed.

Let RI increasing (resp. decreasing) the number of double points be represented as $1a$ (resp. $1b$). For the case when $n = 0$, i.e., $P_0 = P_{HY}$, there exist eight rectangles satisfying the Rule (Figure 7.2). Assuming that $n = k$, k Reidemeister moves are applied to the knot projection, and there exist eight rectangles satisfying the Box Rule. Under this assumption, we consider the case when $n = k+1$. The $(k+1)$th move is $1a$, $1b$, or RⅢ inside the rectangle or is applied outside the rectangles. Note that in this proof, for a Reidemeister move that is *partially* outside the rectangles, we simply say that a Reidemeister move is outside the rectangles.

7.1.4   Moves $1a$, $1b$, and RⅢ inside a rectangle

When RI and RⅢ are applied inside a rectangle, the rectangles for the case when $n = k$ satisfy the Box rule because RI and RⅢ do not change the relation between strands 1 and 2 (condition 1, condition 2).

Condition 3 is implied by condition 1 and condition 2, which is shown as follows. Let F_{j+1} be a region obtained by applying a single Reidemeister move to a $\star$-region when $n = k$. For each rectangle, if strand 3 with no double points exists such that it is a part of the boundary of F_{j+1}, then strand 3 is called a *simple arc 3*. We show that it is not possible for each of two successive rectangles to contain a *simple arc 3* which is contained by the boundary of F_{j+1}. If this is considered to be possible, a *simple arc 3* will be connected to strand 1 or 2, which implies that there is at least one double point outside the rectangles; this contradicts condition 2 that states that there is no double point outside the rectangles. Therefore, it is not possible for each of two successive strands to be a *simple arc 3*. Therefore, the boundary of F_{j+1} has at least four double points in four rectangles. Thus, condition 3 is satisfied.

7.1.5   Moves $1a$, $1b$, and RⅢ outside the rectangles

In this section, we consider moves $1a$, $1b$, or RⅢ outside the rectangles.

7.1.5.1 $1a$

If there exists move $1a$ outside the rectangles, then we can replace a rectangle with a slightly modified rectangle, as shown in Figure 7.3. If this move $1a$ is not completely outside the rectangles, a similar procedure can be followed.

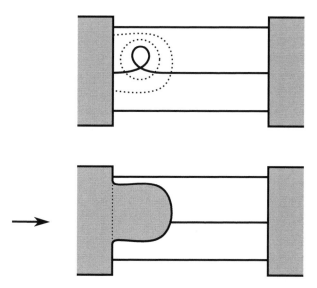

FIGURE 7.3 Deformation for a single $1a$

7.1.5.2 $1b$

Next, we consider that move $1b$ is the $(k + 1)$th move. By the induction assumption, the 1-gon eliminated by $1b$ is placed as the $\star$-region (Case 1) or has sides, which contain the two strands connecting rectangles (Case 2).

However, the $\star$-region has at least four sides and thus cannot be a 1-gon. Therefore, only Case 2 is possible. In this case, consider going along a strand, which is a 1-gon. At least, this strand should be connected with a strand inside a rectangle. If this strand is strand 1 or 2, then it has at least two intersections, which is a contradiction. If the strand is a strand 3 for a rectangle r_1, then this strand should be connected with strand 1 or 2 for another rectangle r_2 that is adjacent to r_1. Therefore, this strand has at least two intersections inside the rectangle, which is a contradiction. As a result, we cannot apply $1b$ as the $(k + 1)$th local move.

7.1.5.3 $RIII$

Finally, we consider a single $RIII$ outside the rectangles. Consider the situation after the kth operation and just before the $(k + 1)$th operation; note that the situation right after the kth operation satisfies the induction assumption. From condition 3 of the Box Rule, $RIII$ cannot be applied in the $\star$-region. From condition 2, every double point of the 3-gon just before the application of the $(k + 1)$th $RIII$ move is inside a box.

Thus, this 3-gon of $RIII$ immediately after $n = k$ can be a 3-gon $\sharp$, as shown in Figure 7.4. If there is a 3-gon of RⅢ, as shown in Figure 7.5, just after

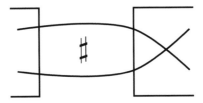

FIGURE 7.4 3-gon ♯ with RⅢ

$n = k$, then a simple arc is placed inside a rectangle; this simple arc should be strand 1 or 2. However, both strands 1 and 2 should have intersections inside a rectangle under the induction assumption. Thus, a 3-gon of RⅢ of type as shown in Figure 7.5 is not possible. Now, we consider the only remaining

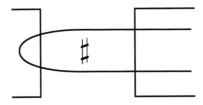

FIGURE 7.5 Impossible 3-gon

possibility. In this case, we can modify the boxes when $n = k$, as shown in Figure 7.6. This operation places a single double point into an adjacent box. In at most two operations, one box can completely contain a 3-gon of RⅢ just before the application of RⅢ as the $(k + 1)$th operation.

As a result, every local move, RI or RⅢ, applied outside the rectangles can be put inside a single rectangle by modifying boxes when $n = k$. Because we have already shown the proof when $n = k+1$ by the induction property for all cases RI and RⅢ, the case when $n = k + 1$ case is established. This completes the proof.

7.2 ARNOLD INVARIANTS

In this section, we review Arnold invariants for plane curves (note that these are not spherical curves). Our exposition of the definition of Arnold invariants may be different from other texts, but, essentially, they should represent the same object. We obtain the definition based on our context.

Before we start introducing Arnold invariants, we recall the common notations used for knot projections on a plane and sphere. To avoid confusion,

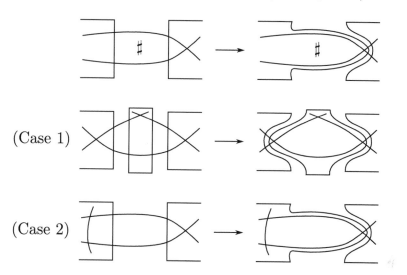

FIGURE 7.6 Covering 3-gon ♯

a knot projection on a plane (resp. sphere) is called a *plane curve* (resp. *knot projection*) and is represented by the symbol C (resp. P) here.

Definition 7.1 A 2-gon on a plane curve or a knot projection belongs to one of the following two types, i.e., strong type or weak type, as shown in Figure 7.7.

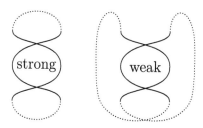

FIGURE 7.7 Strong or weak 2-gon. Dotted curves indicate the connections of the fragments.

Definition 7.2 A 3-gon on a plane curve or a knot projection belongs to one of the following four types, i.e., α, β, γ, and δ, as shown in Figure 7.8. In particular, a 3-gon of type α or β (resp. γ or δ) is called a strong (resp. weak) 3-gon.

Definition 7.3 The second Reidemeister move, RII, can be of two types,

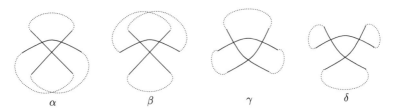

FIGURE 7.8 3-gons of type α, β, γ, and δ. Dotted curves indicate the connections of the sub-curves.

strong or *weak*, depending on the appearance of the *strong* or *weak* 2-gon. A strong RII (resp. weak RII) increasing the number of double points is denoted by $s2a$ (resp. $w2a$). An inverse move is denoted by $s2b$ (resp. $w2b$).

Definition 7.4 The third Reidemeister move, RIII, can be of two types, *strong* or *weak*, depending on the appearance of the *strong* or *weak* 3-gon. A strong RIII (resp. weak RIII) from γ to δ (resp. β to α) is denoted by $s3a$ (resp. $w3a$). An inverse move of $s3a$ (resp. $w3a$) is denoted by $s3b$ (resp. $w3b$).

Definition 7.5 Moves $s2a$, $s3a$, $w2b$, and $w3a$ (resp. $s2b$, $s3b$, $w2b$, and $w3a$) are called *a*-moves (resp. *b*-moves) or *positive* (resp. *negative*) moves.

Remark 7.1 The author feels that the definition of *a*-moves of RI and RII is quite evident to the readers. The above definition of *a*-move for the third Reidemeister moves is also clear.

Recall that chord diagrams are represented by a circle and chords (cf. Chapter 6). Moves $s3a$ and $w3a$ have the direction of increasing the number of sub-chord diagrams of type $\bigotimes$; see Figure 7.9 (Chapter 6 introduced some properties of $\bigotimes$).

Definition 7.6 The sequence of curves K_i is defined in Figure 7.8. For every K_i, the values of J^+, J^-, and St are defined as follows:

$$J^+(K_0) = 0, J^-(K_0) = -1, \text{ and } St(K_0) = 0,$$

$$J^+(K_{i+1}) = -2i, J^-(K_{i+1}) = -3i, \text{ and } St(K_{i+1}) = i.$$

For every plane curve C, we can find a finite sequence s from C to K_i generated by RII and RIII (Chapter 3 also considers K_i). Then, J^+, J^-, and St are defined by the following conditions.

- J^+ increases by 2 under a single $w2a$ and is invariant under strong RII and RIII.

- J^- decreases by 2 under a single $s2a$ and is invariant under weak RII and RIII.

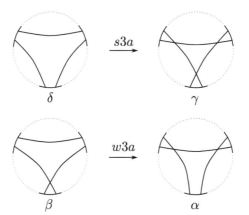

FIGURE 7.9 $s3a$ and $w3a$

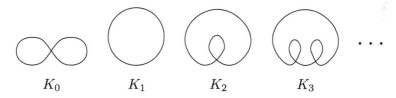

FIGURE 7.10 K_i

- St increases (resp. decreases) by 1 under a single $s3a$ (resp. $w3a$) and is invariant under RII.

It is well known that the definition is well defined, i.e., J^+, J^-, and St do not depend on the path to K_i. Here, we introduce the *averaged invariant* for knot projections on a sphere, defined by Polyak [4]. Recall that to avoid confusion, a knot projection on a plane is called a *plane curve*.

Proposition 7.1 ([4]) *Let P be a knot projection on a sphere and let $C_P(r)$ be a plane curve obtained by arbitrarily placing the infinity point ∞ on a region r from $S^2 \setminus P$. Then, we have the following.*

(1) $a(C_P(r)) = \frac{1}{2}(J^+(C_P(r)) + 2St(C_P(r)))$ does not depend on r. Thus we can set $a(P) = a(C_P(r))$.

(2) $a(P)$ is invariant under RI and strong RII for a knot projection P on a sphere.

Remark 7.2 The average invariant can be represented by a notion that is similar to chord diagrams (cf. [4, 2]).

7.3 EXERCISES

1. See the Gauss diagram formulae by Polyak [4, Theorem 1] and understand that $x(C_P(r))$ does not depend on r for $x = J^+$, J^-, and St.

2. Consider all the possibilities for finding other Arnold-type invariants.

3. Determine why the normalization of Arnold invariants has the setting shown in Definition 7.6.

Bibliography

[1] T. Hagge and J. Yazinski, On the necessity of Reidemeister move 2 for simplifying immersed planar curves. *Knots in Poland III. Part III*, 101–110, Banach Center Publ., 103, *Polish Acd. Sci. Inst. Math., Warsaw,* 2014.

[2] N. Ito and Y. Takimura, Sub-chord diagrams of knot projections, *Houston J. Math.* **41** (2015), 701–725.

[3] O.-P. Östlund, Invariants of knot diagrams and diagramattic knot invariants. Thesis (Ph.D.)–Uppsala Universitet (Sweden). *ProQuest LLC, Ann Arbor, MI,* 2001. 67pp.

[4] M. Polyak, Invariants of curves via Gauss diagrams, Topology **37** (1998), 989–1009.

CHAPTER 8

Further result of strong (1, 3) homotopy

CONTENTS

8.1   Statement 107
8.2   Proof of the statement 108
8.3   Open problems and Exercise 117

M any equivalence classes under strong (1, 3) homotopy exist. Taniyama explicitly described the elements for some of these classes [1]. Not only is this result important but the method is also interesting.

8.1   STATEMENT

Theorem 8.1 ([1]) *Let P_0 and P be knot projections. Let m be a nonnegative integer. Suppose that P_0 has no local disks, (a), (b), (c), or (d), as shown in Figure 8.1. Two knot projections P and P_0 are equivalent under strong (1, 3) homotopy if and only if P is a connected sum of P_0 and knot projections, each of which is a knot projection that appears as ∞ or a trefoil knot projection shown in Figure 8.2.*

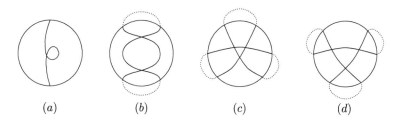

(a)   (b)   (c)   (d)

FIGURE 8.1   Sufficiently small disks (a), (b), (c), and (d)

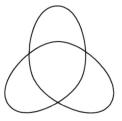

FIGURE 8.2 A trefoil knot projection

8.2 PROOF OF THE STATEMENT

Here, it is sufficient to show Lemma 8.1.

Notation 8.1 *Recall that for two knot projections P_1 and P_2, a connected sum of P_1 and P_2 is denoted by $P_1 \natural P_2$ (cf. Chapter 5). For m ordered knot projections $Q_1, Q_2, \ldots, Q_m$, a connected sum of them is denoted by*

$$(\ldots((Q_1 \natural Q_2) \natural Q_3) \natural \cdots \natural Q_m).$$

Note that each connected sum $\natural$ depends on the arcs, points, and regions (see Chapter 5). A selection of any tuples of arcs, points, and regions gives a connected sum of knot projections, each of which is an element of $\{Q_i\}_{i=1}^m$ and simply denoted by

$$Q_1 \natural Q_2 \natural \cdots Q_m.$$

Lemma 8.1 *Let m be a nonnegative integer. Let P, P_0, and R_i ($1 \leq i \leq m$) be knot projections. If P and P_0 are related by a finite sequence generated by RI and strong RⅢ, then there exists a positive integer m and a set $\{R_i\}_{i=1}^m$ of knot projections such that P is sphere isotopic to*

$$P_0 \natural R_1 \natural R_2 \natural \cdots \natural R_m$$

where each R_i and a trivial knot projection are related by a finite sequence generated by RI and strong RⅢ.

Proof of Lemma 8.1. We show the claim by an induction with respect to the number of applications of RI or strong RⅢ to P_0; here, the number of applied Reidemeister moves is denoted by n. Let P_n be a knot projection obtained from P_0 after the nth Reidemeister move is applied.

• For the case when $n = 0$, there is no R_i; the claim holds.

   We will suppose that the claim is true for case n; we will show the claim under this assumption. For P_n and P_{n+1}, we can consider a sufficiently small disk corresponding to a local replacement that is either RI or strong RⅢ. The sufficiently small disk is called an *x-disk* and is denoted by x_d.

   Now we consider a "fat knot projection" for P_0. More precisely, using a mathematical term, we consider the regular neighborhood B of P_0 (Figure 8.3).

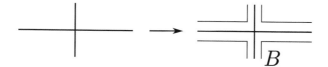

FIGURE 8.3   Regular neighborhood B of knot projection P_0

Recall that if we go along a knot projection, the path starting from a double point to the next double point is called an *arc* (cf. Notation 4.1). Thus, an arc contains one (for 1-gon) or two double points (for n-gon ($n \geq 2$)). In the following, an arc is denoted by a. In the rest of this book, a *simple arc* is a connected part that is a subset of an arc and does not contain any double points. We place a sufficiently small disk r, called a *swelling*, on each arc, where r does not contain any double point of P_0.

FIGURE 8.4   Placing a swelling

A knot projection P_0 without 1- and 2-gons with B and r is shown in Figure 8.5.

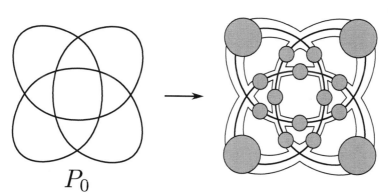

FIGURE 8.5   Knot projection of P_0 with its regular neighborhood ("fat knot projection")

Note that it is clear that the following claims, Claim 8.1 and Claim 8.2, are equivalent.

Claim 8.1 *P is a connected sum* $P_0 \sharp R_1 \sharp R_2 \sharp \ldots \sharp R_m$, *where* R_m *and a trivial knot projection are related by a finite sequence generated by RI and strong R*Ⅲ.

Claim 8.2 *There exists a set of ordered swellings* $\{r_i\}_{i=1}^m$ *such that* r_i *contains* R_i *except for a simple arc, as shown in Figure 8.6.*

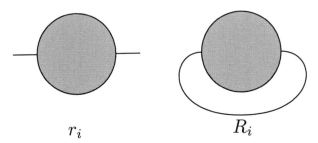

$$r_i \qquad\qquad\qquad R_i$$

FIGURE 8.6 Swelling r_i containing R_i except for a simple arc.

Now suppose that case n for Lemma 8.1 is true, i.e., Claim 8.1 for case n is true. Then, Claim 8.2 for case n holds. Thus, we will show that Claim 8.2 holds for case $(n+1)$ also, which implies Claim 8.1 holds for case $(n+1)$.
• case $(n+1)$. Before considering case $(n+1)$, we will check the property of a swelling.

Suppose that the x-disk x_d contains m double points of P_0, each of which is denoted by $d(P_0)$. Let 1a (resp. 1b) be RI increasing (resp. decreasing) the number of double points. Recall that a single strong RⅢ can be described by 3-gons of types (c) and (d), as shown in Figure 8.1. Let 3a (resp. 3b) be strong RⅢ when switching from (d) to (c) (resp. (c) to (d)) at x_d.

There exist seven cases.

• See Figure 8.7 for the case when $m = 1$.

(1) If 1b is applied for the $(n+1)$th case, x_d contains a single $d(P_0)$ just before the application of 1b (see Figure 8.7 (1)). It is elementary to see that any application of RI or strong RⅢ does not change the (global) connection of a knot projection. Thus, we obtain the center figure of Figure 8.7 (1) (we will explain the reason in the following). Starting at $d(P_0)$, we proceed along P_0 and back to $d(P_0)$. Recall that we suppose that Claim 8.1 holds for case n. If there exist two swellings in this process (going from $d(P_0)$ to $d(P_0)$), at least one other double point of P_0 should appear in x_d of P_n in the left figure in Figure 8.7 (1), which is a contradiction. Thus, as shown in the center figure of Figure 8.7 (1), there exists a single swelling from $d(P_0)$ to $d(P_0)$. Thus, as a part of P_0, this part can be drawn as the right figure, i.e., a 1-gon. However, P_0 has no 1-gons, which is a contradiction. Thus, this case (1) does not occur in x_d.

(2) If $3a$ is applied in the $(n + 1)$th case, x_d will contain a single $d(P_0)$ just before the application of $3a$ (see Figure 8.7 (2)). It is elementary to see that any application of RI or strong RIII does not change the (global) connection of a knot projection. Thus, we have the center figure as shown in Figure 8.7 (2) (we will explain the reason in the following). See this figure. Starting at $d(P_0)$, we proceed along P_0, pass through a swelling, and return to this $d(P_0)$. If there exist two swellings containing two double points appearing in x_d, there exists one additional double point labeled as $d(P_0)$ between the two swellings, which is a contradiction. Thus, we can draw the figure shown in the center figure in Figure 8.7 (2) as a local figure on P_0. Recall the definition of the swellings; hence, for each swelling, we have the following claim (see Figure 8.4).

Claim 8.3 *For a knot projection P and a single swelling r, $P \cap r$ consists of exactly two points.*

Thus, the center figure is converted to the right figure in Figure 8.7 (2). However, there are no 1-gons on P_0, which is a contradiction. Thus, Case (2) does not occur in x_d.

(3) If $3b$ is applied in the $(n + 1)$th case, x_d will contain a single $d(P_0)$ just before the application of $3b$ (see Figure 8.7 (3)). It is elementary to see that any application of RI or strong RIII does not change the (global) connection of a knot projection. Thus, we have the center figure of Figure 8.7 (3) (we will explain the reason in the following). Starting at $d(P_0)$, we proceed along P_0, pass through a swelling, and return to this $d(P_0)$. Two double points, each of which is not $d(P_0)$, appearing in x_d should be contained in a single swelling. This is because if there exists two swellings, each of which contains a double point in x_d, then there exists one additional double point labeled as $d(P_0)$ between the two double points. However, such a double point does not appear in x_d. Thus, we have the figure as shown in the center of Figure 8.7 (3). Focus on a single swelling and the double point labeled as $d(P_0)$. From Claim 8.3, we know that there exists at least one 1-gon as shown in the right figure of Figure 8.7 (3). However, P_0 contains no 1-gons, which is a contradiction. Thus, Case (3) does not occur in x_d.

- Figure 8.8 showns the case when $m = 2$. Suppose that two double points labeled as $d(P_0)$ appear in x_d. Then, there exist two cases (4) and (5) as shown below.

(4) If $3a$ is applied in the $(n+1)$th case, x_d will contain two double points, each of which is labeled as $d(P_0)$, just before the application of $3a$ (as shown in the left figure, Figure 8.8 (4)). From this figure, P_0 should be as shown in the center figure of Figure 8.8 (4) (we will explain the reason in the following). It is elementary to see that any application of RI or strong RIII does not change the (global) connection of a knot projection.

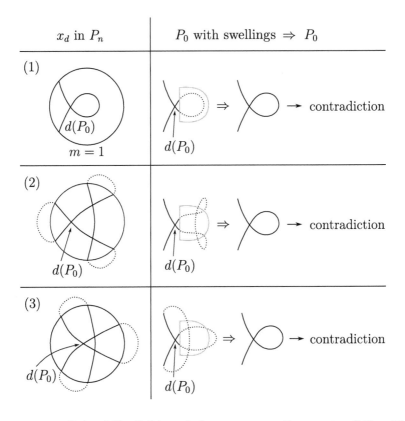

x-disk of P_n (left) and the corresponding parts of P_0 with a swelling (center) and without swellings (right)

An unlabeled double point d should be in a swelling s. Starting at d, we proceed along P_n and return to d avoiding the other two double points in x_d. Claim 8.3 requires an entire dotted curve to be included in the swelling s. Again, according to Claim 8.3, there exists another swelling between two double points labeled as $d(P_0)$. Thus, we have the figure, which is shown as the center figure Figure 8.8 (4) representing P_0. Therefore, P_0 can be represented as shown in the right figure of Figure 8.8 (4). However, there is no 2-gon of type (b), as shown in Figure 8.1, which is a contradiction.

(5) If $3b$ is applied in the $(n+1)$th case, x_d will contain two double points, each of which is labeled as $d(P_0)$, just before the application of $3b$ (the left figure of Figure 8.8 (5)). Focus on the two double points, each of which is labeled as $d(P_0)$. Claim 8.3 requires the existence of a swelling that contains an unlabeled double point d, as shown in the upper center figure of Figure 8.8 (5). From Claim 8.3, we know that there are no double points on dotted curve α; and then we can extend the swelling containing unlabelled double point d until the swelling contains the entire α. We apply similar argument to the dotted curve β. In conclusion, both α and β are simple arcs in P_0. In this situation, there exists a 3-gon of type (c). However, there is no 3-gon of type (c) in P_0, which is a contradiction.

- The case when $m = 3$ is discussed below.

(6) If $3a$ is applied in the $(n + 1)$th case, x_d will contain a three $d(P_0)$ just before the application of $3a$. By the induction assumption for the nth case, there exists an integer m and a set $\{R_i\}_{i=1}^{m}$ such that P_n is represented as

$$P_0 \natural R_1 \natural R_2 \natural R_3 \natural \ldots \natural R_m.$$

Thus, there exists a disk, as shown in the center figure in Figure 8.9 (6), containing three double points, each of which is labeled as $d(P_0)$. Thus, there exists a 3-gon of type (d) in P_0, which is a contradiction.

(7) If $3b$ is applied in the $(n + 1)$th case, x_d will contain three double points, each of which is labeled as $d(P_0)$ just before the application of $3b$. By the induction assumption for the nth case, there exists an integer m such that P_n is a connected sum $P_0 \natural R_1 \natural R_2 \natural \cdots \natural R_m$ for $\{R_i\}_{i=1}^{m}$. Thus, P_0 has a local disk corresponding to x_d and the disk contains a 3-gon of type (c) (Figure 8.9 (7)). However, P_0 contains no 3-gons of type (c), which is a contradiction.

As a result, there is no double point in x_d. Therefore,

$$x_d \subset \bigcup_{1 \leq i \leq m} r_i \cup B^c$$

where $B^c = S^2 \setminus B$.

Further, we have Lemma 8.2.

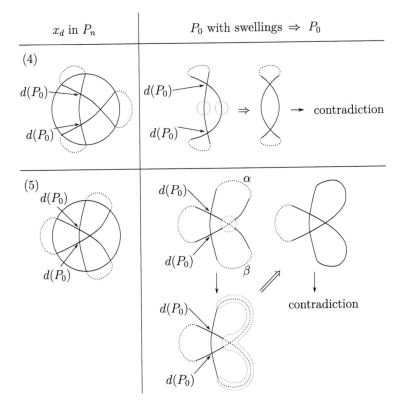

FIGURE 8.8 x-disk of P_n (left) and the corresponding parts of P_0 with swellings (center) and without swellings (right)

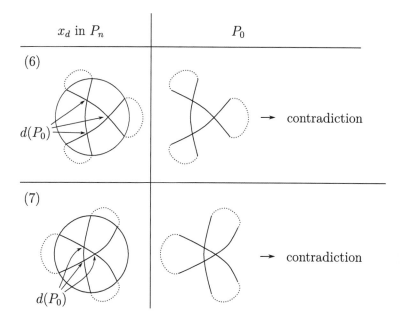

FIGURE 8.9 x-disk of P_n (left) and the corresponding part of P_0 (right)

Lemma 8.2 x_d *cannot contain two double points such that one belongs to* r_i *and the other belongs to* r_j $(i \neq j)$.

Proof of Lemma 8.2. It is sufficient to verify the statement for the case where there exists a 3-gon in x_d. In this case, as shown in Figure 8.10, there exists r_i and r_j such that r_i (resp. r_j) contains a double point (resp. another double point) in x_d.

By the induction assumption for the nth case, for any r_i and r_j, if we proceed along P_n starting from r_i, we have encountered at least one double point before we encounter r_j first. This is because any two neighboring swellings are as shown in Figure 8.11. Therefore, the existence of two double points in x_d such that one belongs to r_i and the other belongs to r_j $(i \neq j)$ is not possible. □

Note that we have Lemma 8.2. Thus, by redefining a sufficiently small disk as x_d if required, it is possible to have at most one swelling that intersects x_d. Now, for a swelling r_0, we have

$$x_d \subset r_0 \cup B^c.$$

Recall that for the nth case, we have Claim 8.2. Thus, for the nth case, $P_n \subset \bigcup_{i=1}^m r_i \cup B$. Then, if required, we can redefine x_d such that it is sufficiently small, i.e., $x_d \subset r_0$, just before the application of the $(n+1)$th Reidemeister move for the nth case. Thus, Claim 8.2 holds for the $(n+1)$th

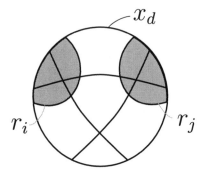

FIGURE 8.10 x_d with r_i and r_j

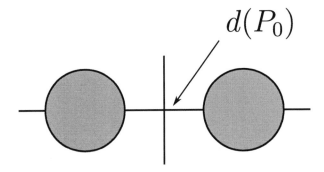

FIGURE 8.11   Two neighboring swellings

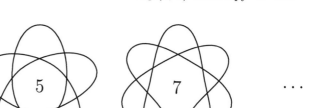

FIGURE 8.12 Infinite family of knot projections

case, which implies Claim 8.1 for the $(n + 1)$th case. By induction, we have Claim 8.1.

Here, recall the fact shown in Chapter 5.

Fact 8.1 *Let P be a knot projection and $\bigcirc$ be a trivial knot projection. The following two conditions, (1) and (2), are equivalent:*
(1) P and $\bigcirc$ are related by a finite sequence generated by RI and strong RIII.
(2) P is a connected sum of knot projections, each of which is a trivial knot projection, a knot projection that appears as ∞, and a trefoil knot projection, as shown in Figure 8.2.

From Claim 8.1, for any i $(1 \leq i \leq m)$, R_i and $\bigcirc$ are related by a finite sequence generated by RI and strong RIII. Thus, for R_i corresponding to r_i, we apply Fact 8.1 to R_i. Hence, we have the statement of Theorem 8.1 and we complete the proof.

8.3 OPEN PROBLEMS AND EXERCISE

1. Find non-trivial knot projections satisfying the condition in Theorem 8.1 for P_0 without using a connected sum. For example, the following knot projections (shown in Figure 8.12) satisfy the assumption made in Theorem 8.1.

2. (open) Determine other equivalence classes under strong (1, 3) homotopy, i.e., an equivalence class not satisfying the condition of the statement in Theorem 8.1.

3. (open, except for one example) Find an invariant characterizing an equivalence class under strong (1, 3) homotopy. For example (only one example), for the equivalence class containing a trivial knot projection, we can have such an invariant. That is

$$3 \oplus (P) - 3 \otimes (P) + \otimes (P) = 0$$

is a necessary and sufficient condition for knot projection P and a trivial knot projection ◯ to be related by a finite sequence generated by RI and strong RⅢ.

Bibliography

[1] N. Ito, Y. Takimura, and K. Taniyama, Strong and weak (1, 3) homotopies on knot projections, *Osaka J. Math.* **52** (2015), 617–647.

CHAPTER 9

Half-twisted splice operations, reductivities, unavoidable sets, triple chords, and strong $(1, 2)$ homotopy

CONTENTS

9.1   Splices 122
9.2   Ito-Shimizu's theorem for half-twisted splice operations .. 124
9.3   Unavoidable sets 125
9.4   Upper bounds of reductivities 130
9.5   Knot projection with reductivity one: revisited 138
9.6   Knot projection with reductivity two 146
9.7   Tips 150
     9.7.1   Tip I 150
     9.7.2   Tip II 150
9.8   Open problems and Exercises 151

Half-twisted splice is the inverse of a splice of a double point of a knot projection for obtaining a knot projection from another knot projection. A "splice" is a replacement of a sufficiently small disk with another disk while preserving its boundary on a knot projection; here, a local part of two paths at a double point is replaced with two sub-curves with no double points. Splice is considered to be the basic tool in knot theory. There are two types of splices, one is called a *Seifert splice* and the other is the inverse of a half-twisted splice, denoted by A^{-1} [1]. In this study, we suppose that every knot projection has at least one double point. For a given knot projection, the notion of the inverse of the half-twisted splice A^{-1} induces Shimizu's reductivity $r(P)$ that is the minimum number of applications of A^{-1}, recursively, to obtain a reducible knot projection from P. The reductivity is closely related to an open question concerned with an unavoidable set. Roughly speaking, an unavoidable set is a set for which the following condition holds: any knot projection must contain a part that is sphere isotopic to one of the elements. If a knot projection contains a single 3-gon, the types of 3-gons can be easily classified into four types. Two of them that appear on two sides of a weak RⅢ are denoted by A and B; the other two that appear on the two sides of a strong RⅢ are denoted by C and D. The open question is that for a reduced knot projection, is {2-gon, 3-gons of type A, B, and C} an unavoidable set? This open question concerned with an unavoidable set is also related to the triple chord theorem [3], which is stated as follows: the chord diagram of any prime knot projection with no 1-gons and 2-gons contains triple chords.

In this chapter, we suppose that every knot projection has at least one double point.

9.1 SPLICES

In 2012, a *half-twisted splice* and its inverse were introduced [1]. In this chapter, the term *local replacement* represents a replacement of a sufficiently small disk such as that shown in the left figure in Figure 9.1. *Splice* is the local replacement of a disk d with another disk d' that contains two simple arcs a, a' such that $\partial d = \partial d'$ (i.e., the boundary is fixed) and $d \cap P = d' \cap (a \cup a')$ (i.e., the four points are fixed), as shown in Figure 9.1 (for the definition of simples arcs, see Chapter 8, just before Claim 8.1). Any splice belongs to one of two types of splices mentioned earlier. To define the type of splice, as one of the ways, we use the orientation. If a knot projection is arbitrarily orientated, then the inverse of a half-twisted splice (resp. Seifert splice) is a splice defined by Figure 9.2 (resp. Figure 9.3).

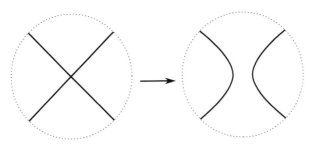

FIGURE 9.1 Local replacement

A *nugatory double point* is a type of double point, p, shown in Figure 9.4. That is, if we apply Seifert splice to p, there exists a simple closed curve that

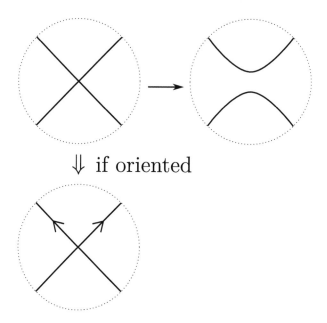

FIGURE 9.2 Inverse of a half-twisted splice

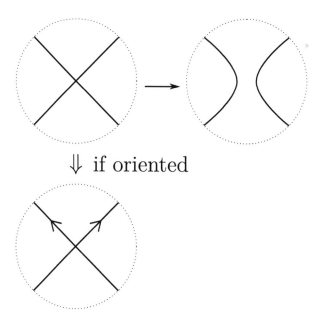

FIGURE 9.3 Seifert splice

coincides with the boundary of a disk containing exactly one knot projection; then, p is called a *nugatory double point*.

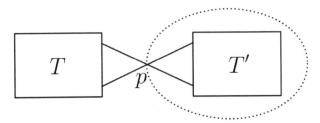

FIGURE 9.4 Nugatory double point p

A knot projection P is *reducible* if there exists a double point p of P such that p is a nugatory double point. A knot projection P is said to be reduced if P is not reducible.

9.2 ITO-SHIMIZU'S THEOREM FOR HALF-TWISTED SPLICE OPERATIONS

A trefoil projection is (a sphere isotopy class of) a knot projection as shown in Figure 9.5.

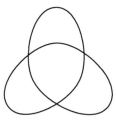

FIGURE 9.5 A trefoil projection

Theorem 9.1 ([1]) *Let P be a reduced knot projection. Let A (resp. A^{-1}) be a half-twisted splice (resp. the inverse of a half-twisted splice), which is a local replacement shown in Figure 9.6. Then there exists a finite sequence generated by A and A^{-1} from P to a trefoil projection.*

This proof is slightly technical and thus we omit the proof here (however, the required preliminary knowledge is not much, see [1]). However, the result leads to the following question.

Question 9.1 *How do we count the minimum number of applications of A^{-1} to obtain a reducible knot projection?*

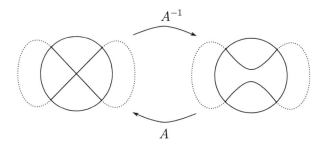

FIGURE 9.6 A half-twisted splice A^{-1} and its inverse A at a double point of a knot projection P in a disk (the dotted sub-curves indicate the connections of P outside the disk)

The minimum number, as stated in Question 9.3, can be used to obtain a classification of the knot projections as follows.

The reductivity $r(P)$ of a knot projection P was introduced by Shimizu [4] (cf. Chapter 4).

Definition 9.1 Let P be a reduced knot projection. $r(P) = \min\{$number of inverses of half-twisted splices, applied recursively, to obtain a reducible knot projection from $P\}$.

Shimizu [4] asked the following question:

Question 9.2 *Is it true that for every knot projection P, $r(P) \le 3$?*

9.3 UNAVOIDABLE SETS

This reductivity is closely related to the notion of an *unavoidable set* for the knot projections [4].

Definition 9.2 ((n,n)-tangle) Let n be a positive integer. A subset of $\mathbb{R}$ that is plane isotopic to $[0,1]$ is called an *interval*. An (n,n)-*tangle* is the image of a generic immersion of n intervals and a finite number of circles into $\mathbb{R} \times [0,1]$ such that the boundary points of the intervals map bijectively to the $2n$ points

$$\{1,2,\ldots,2n\} \times \{0\}, \{1,2,\ldots,2n\} \times \{1\}.$$

These $2n$ points are called the *endpoints of a tangle*. For an (n,n)-tangle, if n is not specified, we simply call it a *tangle*.

Let T be a tangle on $\mathbb{R}^2$. If there exists an embedding ι from $\mathbb{R}^2$ to S^2, $\iota(T)$ is called a *tangle on a sphere*.

Definition 9.3 (unavoidable set) Let n be a positive integer. Consider a

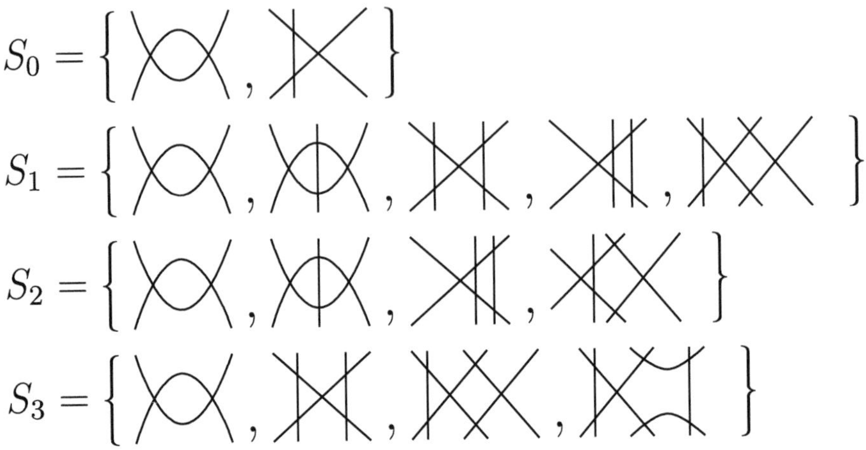

$$S_0 = \left\{ \ \text{} \ \right\}$$

$$S_1 = \left\{ \ \text{} \ \right\}$$

$$S_2 = \left\{ \ \text{} \ \right\}$$

$$S_3 = \left\{ \ \text{} \ \right\}$$

FIGURE 9.7 Unavoidable sets

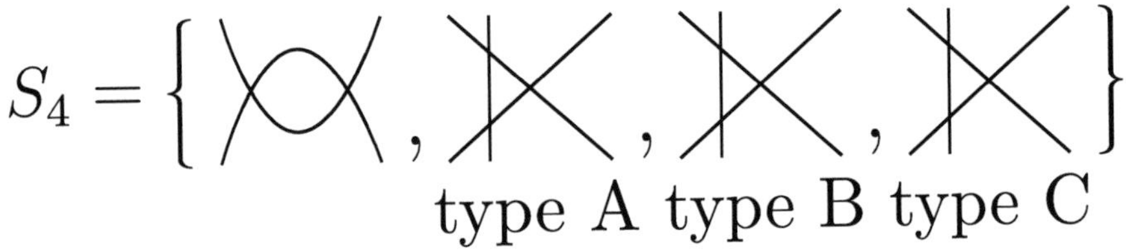

$$S_4 = \left\{ \ \text{} \ \right\}$$

type A type B type C

FIGURE 9.8 S_4

set S consisting of tangles included in a knot projection. If every knot projection contains at least one element of S, then S is called an *unavoidable set*. If S is not an unavoidable set, it is an *avoidable set*.

Theorem 9.2 ([4]) *The sets S_0, S_1, S_2, and S_3 shown in Figure 9.7 are unavoidable sets.*

However, the answer to the following question is not known.

Question 9.3 *Is it true that S_4 is an unavoidable set?*

Here, 3-gons of types A, B, C, and D are of types α, β, γ, and δ, respectively, as shown in Figure 9.9 (recall that every 3-gon of a knot projection belongs to one of the four types, cf. Chapter 7 (Definition 7.2)) where the dotted subcurves indicate the connections of the endpoints of each tangle.

Proof of Theorem 9.2. Let P be a reduced knot projection. Let V be the number of double points, E be the number of edges, and F be the number of faces. Here, an *edge* is an arc of a knot projection (see Notation 4.1). A *face*

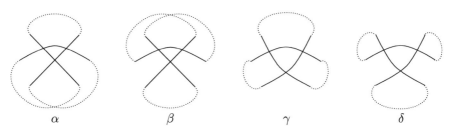

FIGURE 9.9   3-gons of types α, β, γ, and δ

is a disk whose boundary consists of a finite number of edges. If we know that the number of the Euler characteristic number of a sphere is 2, then

$$V - E + F = 2. \tag{9.1}$$

This proof is well-known (refer to see Section 9.7). For every knot projection with no 1-gons, the boundaries of exactly four edges have a single common double point. The boundary of a single edge has exactly two double points. Thus,

$$4V = 2E. \tag{9.2}$$

Let C_k be the number of k-gons in P. Then,

$$F = \sum_k C_k, \tag{9.3}$$

$$2E = \sum_k kC_k. \tag{9.4}$$

If P is reduced, then P has no 1-gons. Using the above four formulae, i.e., (9.1), (9.2), (9.3), and (9.4), we have

$$\begin{aligned}
8 &= 4V - 4E + 4F \\
&= -2E + 4F \\
&= \sum_k (4 - k)C_k \\
&= 2C_2 + C_3 + \sum_{k \geq 4} (4 - k)C_k.
\end{aligned} \tag{9.5}$$

We show S_0, S_1, S_2, and S_3 are unavoidable sets as follows.

• S_0: Because $\sum_{k \geq 4} (4 - k)C_k \leq 0$, S_0 is an unavoidable set.

• S_1: Assume that a rational number is assigned to every k-gon and each rational number is called a *charge*. We set the charge of every k-gon to be $(4 - k)$.

Suppose that there are no 2-gons. Then there exist at least eight 3-gons (see (9.5)). For every 3-gon T, we consider the following operations (see Figure 9.10); this process is referred to as *discharging*.

1. For a face S sharing a common edge with T, $1/6$ is added to the charge of S and $1/6$ is discharged from T.

2. For a face R sharing a common vertex with T, $1/6$ is added to the charge of R and $1/6$ is discharged from T.

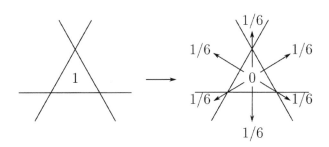

FIGURE 9.10 Discharging from 3-gon T.

Suppose S_1 is an avoidable set, i.e., there exists a knot projection having no element of S_1. After discharging, the following condition is satisfied.

1. The charge of a 3-gon is 0.

2. By the definition of S_1, the charge of a 4-gon will not be changed.

3. If there exists a 5-gon F_5 and m 3-gons have common edges with F_5, the charge of F_5 is
$$-1 + m/6.$$
Thus, if $-1 + m/6 \geq 0$, then
$$m \geq 6.$$

Now there are no 1-gons and 2-gons. Then, because $m \geq 6$, there exists at least six faces that have edges in common with those of F_5. However, by the definition of S_1, this is a contradiction. Thus, either the charge of F_5 is negative or there exist no 5-gons.

4. Let n be a positive integer greater than 5. Suppose there exists an n-gon F_n such that the 3-gons have m edges in common with F_n. The charge of F_n is
$$4 - n + m/6.$$
Here, if $4 - n + m/6 \geq 0$, then
$$m \geq 6(n - 4) \geq 2n + 4(n - 6) \geq 2n.$$

Then, F_n shares n vertices and n edges with 3-gons. This contradicts with the definition of S_1.

As a result, the assumption that "S_1 is an avoidable set" is false, and S_1 is an unavoidable set.

- S_2. The proof is similar to that of S_1; however, we repeat several steps to help the readers (please compare the proofs for S_2 and S_1). Note that the terminology used for *charge* $(4-k)$ is the same as that used for the case S_1. The difference lies in the *discharging* process.

Suppose that there are no 2-gons. Then there exist at least eight 3-gons. For every 3-gon T, we consider the following operation (see Figure 9.11).

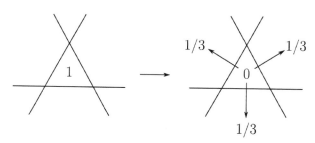

FIGURE 9.11 Discharging from 3-gon T.

- (operation). For a face S sharing the common edge with T, $1/3$ is added to the charge of S and $1/3$ is discharged from T.

Suppose S_2 is an avoidable set. After the discharging process, the following conditions are satisfied.

1. The charge of 3-gon is 0.

2. By the definition of S_2, a 4-gon should not be charged, i.e., its charge is 0.

3. By the definition of S_2, a 5-gon should not be charged, i.e., its charge is -1.

4. Let n be a positive integer greater than 5. Suppose that there exists an n-gon F_n such that F_n shares m edges with 3-gons. The charge of F_n is

$$4 - n + m/3.$$

Here, if $4 - n + m/3 > 0$, then

$$m > 3(n-4) \geq n + 2(n-6) \geq n.$$

Thus, there are m $(\geq n+1)$ edges in F_n. However, it is impossible for F_n to have m $(\geq n+1)$ edges; this is a contradiction.

As a result, the assumption that "S_2 is an avoidable set" is false, and S_2 is an unavoidable set.

• S_3. The proof for S_3 is the same as that of S_2 except for the process of discharging.

Suppose that there are no 2-gons. Then there exists at least eight 3-gons. For every 3-gon T, we consider the following operations (Figure 9.12), called *discharging*.

• (operation). For a face R sharing a common vertex with T, 1/3 is added to the charge of R and 1/3 is discharged from T.

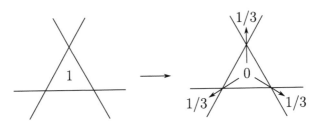

FIGURE 9.12 Discharging from 3-gon T

Suppose S_3 is an avoidable set. After the discharging process, the following conditions are satisfied.

1. The charge for each 3-gon is 0.

2. The charge for each 4-gon, from the definition of S_3, is 0.

3. The charge for each 5-gon, by the definition of S_3, is -1.

4. Let n be a positive integer greater than 5. Suppose that there exists an n-gon F_n such that F_n shares m vertexes with 3-gons. The charge of F_n is

$$4 - n + m/3.$$

Here, if $4 - n + m/3 > 0$, then

$$m > 3(n - 4) = n + 2(n - 6) \geq n.$$

Then, F_n has at least m ($\geq n + 1$) edges, which is a contradiction.

As a result, the assumption that "S_3 is an avoidable set" is false, and hence S_3 is an unavoidable set.

9.4   UPPER BOUNDS OF REDUCTIVITIES

In the following, 3-gons of types A, B, C, and D, and strong and weak 2-gons are defined in Figure 9.13. In this section, we review the results of [4] and [2] for upper bounds of reductivities r, t, y, and i.

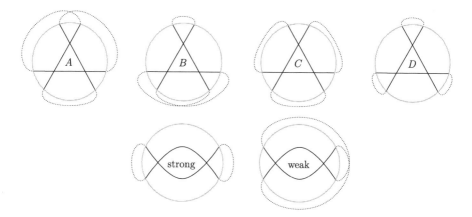

FIGURE 9.13 Dotted fragments indicate the connections outside the disks

Definition 9.4 Let P be a knot projection. Nonnegative integers $t(P)$, $y(P)$, and $i(P)$ are defined as follows.

- $t(P) = \{$number of splices applied simultaneously to obtain a reducible knot projection from $P\}$.

- $y(P) = \{$number of inverses of half-twisted splices applied simultaneously to obtain a reducible knot projection from $P\}$.

- $i(P) = \{$number of Seifert splices applied simultaneously to obtain a reducible knot projection from $P\}$.

Lemma 9.1 *Let P be a knot projection. If there exists a simple closed curve intersecting P at least four points and the connections of P are as shown in the left figure of Figure 9.14. Then, there exists a double point d such that if we apply A^{-1} to d, we have another knot projection P', which is as shown in the right figure of Figure 9.14.*

Proof of Lemma 9.1. Refer to Figure 9.15. Two double points, selected alternatively from four points on the circle, are denoted by a (resp. x) and b (resp. y) as shown in Figure 9.15. Now we choose an orientation for P, as shown in Figure 9.15. A point $a = a(t)$ starts moving from point $a(0)$ and proceeds along P until $a(s)$ arrives at x and another moving point $b = b(s)$ start moving from point $b(0)$ and proceeds along P until $b(t)$ arrives at y, where s and t are parameters. Then, we consider two possibilities as follows. The double point d consists of two local paths, where one path that is passed by a is denoted by (the same symbol) a and the other path that is passed by b is denoted by (the same symbol) b. Case (1) is when the orientation from path a to path b is clockwise and Case (2) is when it is counterclockwise.

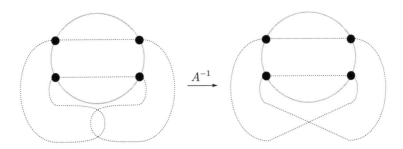

FIGURE 9.14 A^{-1} from P to P' (the dotted fragments indicate the connections of P)

When we apply A^{-1} to double point d, we have the claim for each case, as shown in the lower line of Figure 9.15. □

• Upper bounds concerned with a weak 2-gon.

Refer to Figure 9.16. If a knot projection P has at least one weak 2-gon, then we can easily find a double point such that if we apply A^{-1} to the double point, we will obtain a reducible knot projection. Thus $r(P) \leq 1$.

• Upper bound concerned with a strong 2-gon.

Refer to Figure 9.16. If a knot projection P has at least one strong 2-gon, then we can find a circle surrounded by the strong 2-gon, as shown in Figure 9.17; we can apply Lemma 9.1 to the circle. After applying A^{-1} from Lemma 9.1 to the circle, we obtain a weak 2-gon. Thus, $t(P) \leq r(P) \leq 2 = 1 + 1$.

• Upper bound concerned with a 3-gon of type A.

If a knot projection P has at least one 3-gon of type A, then it is easy to see $y(P) \leq 2$ by finding two vertices, a and b, of the 3-gon on which A^{-1} is applied twice. It also can be seen that if we apply A^{-1} to a, then we have a weak 2-gon. Thus, $r(P) \leq 1 + 1 = 2$.

• Upper bound concerned with a 3-gon of type B.

If a knot projection has at least one 3-gon of type B, it can be seen that if we apply a vertex of the 3-gon, we obtain a strong 2-gon and then $t(P) \leq r(P) \leq 1 + 2 = 3$.

• Upper bound concerned with a 3-gon of type C.

Suppose that a knot projection P has at least one 3-gon of type C. If there exists a simple closed curve γ, as shown in Figure 9.20, and one of the three dotted sub-curves is completely contained inside a disk with boundary γ, then it is elementary to show that $r(P) \leq 1$. If not, two dotted intersecting sub-curves exist as shown in Figure 9.21. In this case, by applying Lemma 9.1 to the outside of the circle, as shown in the left figure of Figure 9.21, there exists a double point on which A^{-1} can be applied to obtain a knot projection with a 3-gon of type A. Thus, $t(P) \leq r(P) \leq 1 + 2 = 3$.

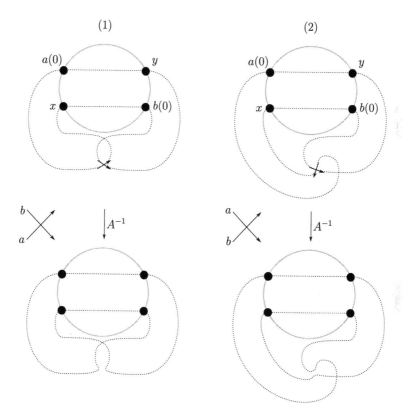

FIGURE 9.15 Two cases of the application of A^{-1}

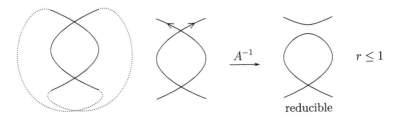

FIGURE 9.16 If P has a weak 2-gon, $r(P) \le 1$.

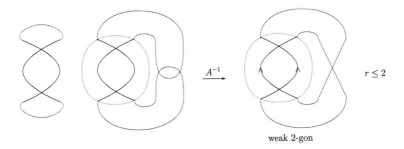

FIGURE 9.17 If P has a strong 2-gon, $r(P) \le 2$.

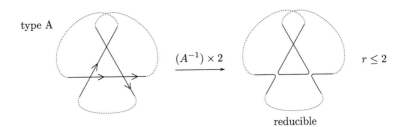

FIGURE 9.18 A knot projection having a 3-gon of type A

Further, we can find two double points such that if Seifert splices are applied to them, then a reduced knot projection is obtained, as shown in Figure 9.30, which implies $t(P) \le i(P) \le 2$.

- Upper bound concerned with a 3-gon of type D.

- Suppose that a knot projection has a 3-gon of type D. If there exists a simple closed curve, as shown in Figure 9.22, then it can be easily seen that $r(P) \le 1$.

If not, there exist two dotted sub-curves such that they intersect with each other, as shown in Figure 9.23. By applying Lemma 9.1 to the part that is

type B

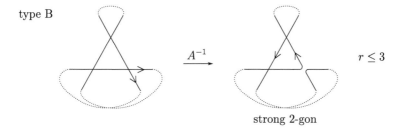

strong 2-gon

FIGURE 9.19 A knot projection having a 3-gon of type B

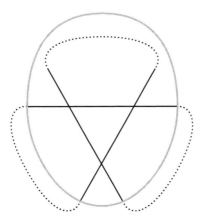

FIGURE 9.20 A knot projection having a 3-gon of type C, dotted arcs indicate the connections of the fragments

type C

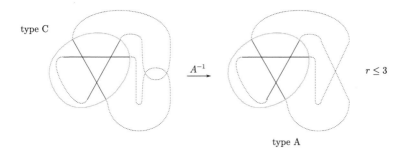

type A

FIGURE 9.21 A knot projection having a 3-gon of type C

outside a simple closed curve, as shown in the left figure of Figure 9.23, we obtain a 3-gon of type B; then, $t(P) \leq r(P) \leq 1 + 3 = 4$.

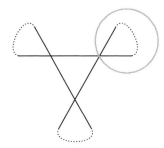

FIGURE 9.22 A knot projection having a 3-gon of type D

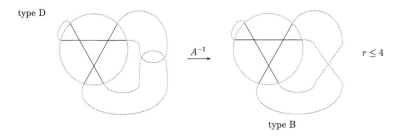

FIGURE 9.23 A knot projection having a 3-gon of type D

Finally, we show that if a knot projection P has a 3-gon of type B, $y(P) \leq 3$. Before proving this statement, we show Lemma 9.2.

Lemma 9.2 *Let P be a knot projection. If there exists a simple closed curve intersecting P at least four points and the connections of P are as shown in the left figure in Figure 9.24, then there exists a double point d such that if A^{-1} is applied to it, we will obtain another knot projection P', as shown in the right figure of Figure 9.24.*

Proof of Lemma 9.2. Refer to Figure 9.25. The proof for this is the same as that of Lemma 9.1 except for the figure shown in Figure 9.25. Thus, we omit describing the same discussion here. □

Now we prove $y(P) \leq 3$ if P has a 3-gon of type B. Refer to Figure 9.26. First, we consider the case such that there exists a simple closed curve intersecting P at just a single double point (Figure 9.26, left). In this case, P is reducible and hence $y(P) = 0$. Second, if such a simple closed curve does not exist, for the center figure of Figure 9.26, by symmetry, we can assume that sub-curve a intersects sub-curve b. By Lemma 9.2, with respect to a simple closed curve of the center figure of Figure 9.26, we obtain $y(P) \leq 3$.

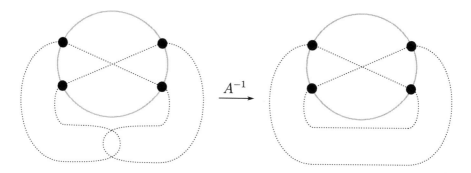

FIGURE 9.24 An application of A^{-1} from P to P' (the dotted fragments indicate the connections of a knot projection)

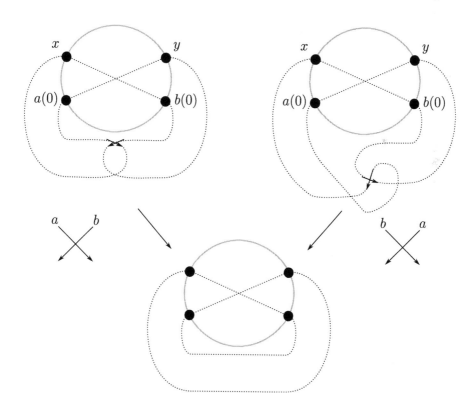

FIGURE 9.25 Two cases of the application of A^{-1}

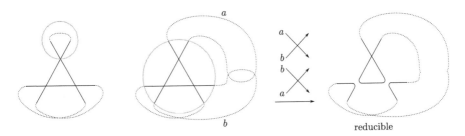

FIGURE 9.26 Knot projection having a 3-gon of type B satisfying $y(P) \leq$ 3 for Case 1 (left) and Case 2 (center, right)

9.5 KNOT PROJECTION WITH REDUCTIVITY ONE: REVISITED

In this section, we recall the necessary and sufficient condition is that Shimizu's reductivity $r(P)$ should be one (see Chapter 4) [3].

Theorem 9.3 ([3] (cf. Chapter 4)) *Let P be a knot projection and $r(P)$ be the reductivity of P defined as above. $r(P) = 1$ if and only if there exists a circle intersecting P at exactly two double points of P, as shown in Figure 9.27.*

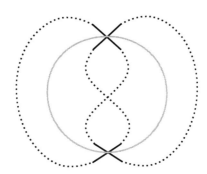

FIGURE 9.27 Knot projection with a simple closed curve

Moreover, the latter condition is equivalent to $t(P) = y(P) = 1$ for reductivities t and y.

From the definitions of reductivities r, t, y, and i, we have Proposition 9.1.

Proposition 9.1 ([2])   1. *For any knot projection P, $t(P) \leq r(P)$, $y(P)$, $i(P)$.*

2. *For any knot projection P, $i(P) \in 2\mathbb{Z}_{\geq 0}$. For a given nonnegative integer $2m$, there exists a knot projection Q such that $i(Q) = 2m$.*

3. *For any reduced knot projection P having at least one strong 2-gon, $i(P) = 2$.*

4. *For any reduced knot projection P having at least one 3-gon of type C, $i(P) = 2$.*

5. *For any reduced knot projection P having at least one 3-gon of type A, $1 \leq y(P) \leq 2$.*

Proof of Proposition 9.1.

1. The condition of splices in the definition of $t(P)$ is weaker than those of the other reductivities; that is, if a reductivity is less than or equal to s, then $t(P) \leq s$, which is obtained from the condition of the splices of $t(P)$.

2. A single Seifert splice changes a component of a (multi-component) curve. Thus, at least two Seifert splices are needed to obtain a knot projection from another knot projection.

   Next, refer to Figure 9.28. For a given nonnegative integer m, a knot projection with $2m + 1$ double points can be obtained from Figure 9.28 and is called a $(2, 2m + 1)$-*torus knot projection*. The $(2, 2m + 1)$-torus knot projection P_m has $i(P_m) = 2m$.

3. Figure 9.29 implies the claim. When a given knot projection is reduced, then the two distinct dotted sub-curves intersect and at least one double point d is obtained. We first apply Seifert splice to a single double point on the boundary of the strong 2-gon. Then, we have at least one 1-gon in one of the two components of the resultant curve C. We can place a simple closed curve intersecting C at just one double point. We apply a single Seifert splice to d and we obtain a knot projection with 1-gon, which is reducible.

4. Figure 9.30 implies the claim.

5. Figure 9.31 implies the claim.

□

Now, we will state the necessary and sufficient condition i.e., $t(P) = 1$, as shown in Theorem 9.4 (cf. Theorem 9.3, its proof is shown in Chapter 4). Theorem 9.4 states that the two conditions $(\star)$ and $(\star\star)$ are equivalent for a reduced knot projection P.

Theorem 9.4 ([2]) *The following two conditions $(\star)$ and $(\star\star)$ are equivalent.*
$(\star)$ *There exists a simple closed curve that intersects at just two double points of P, as shown in Figure 9.27.*
$(\star\star)$ $t(P) = 1$.

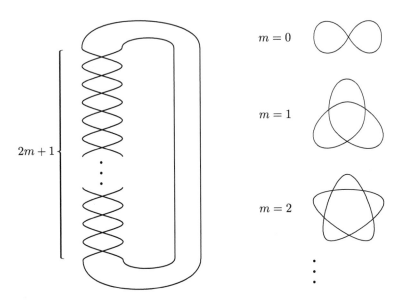

$m = 0$

$m = 1$

$m = 2$

$2m + 1$

FIGURE 9.28 $(2, 2m + 1)$-torus knot projection

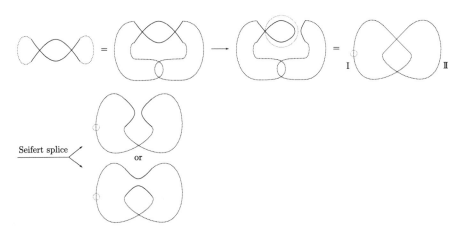

Seifert splice

or

FIGURE 9.29 P with a strong 2-gon implying $i(P) = 2$

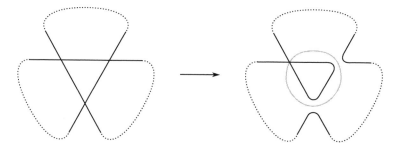

FIGURE 9.30   3-gon of type C implying $i(P) = 2$

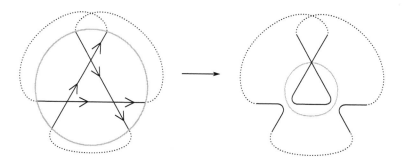

FIGURE 9.31   3-gon of type A implying $y(P) \leq 2$.

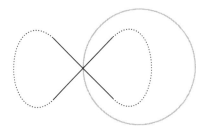

FIGURE 9.32 Simple closed curve and nugatory double point

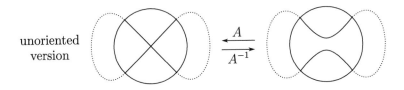

FIGURE 9.33 Half-twisted splice operation (unoriented version)

It is easy to see that $(\star\star)$ implies $(\star)$. Thus, we show that "$(\star)$ implies $(\star\star)$" from (Step 1)–(Step 3).

(Step 1) Considering the definition of a reducible knot projection, we focus on a circle that divides S^2 into two disks and intersects at just one nugatory double point (Figure 9.32).

(Step 2) Consider every possibility of an inverse operation of a splice on the circle. Note that a single Seifert splice does not modify a knot projection to another knot projection (because Seifert splice changes the components of a curve). Thus, we consider only the possibility that the corresponding splice is the inverse of a half-twisted splice. Note that the definitions of a half-twisted splice and inverse operations are obtained for unoriented knot projections (Figure 9.33). However, to omit the cases that are not possible efficiently, orientations can be used (Figure 9.34). From this, it is easy to see possible orientations in Figure 9.35.

(Step 3) Just before applying the inverse operation of a half-twisted splice, we choose a knot projection from the possibilities (Figure 9.35). □

As a corollary, we have

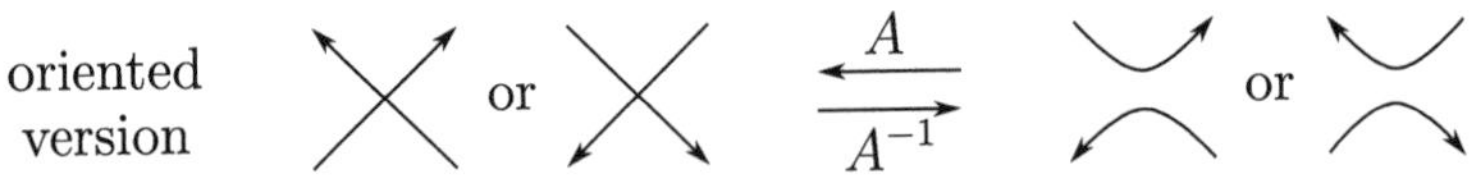

FIGURE 9.34 Half-twisted splice operation (oriented version)

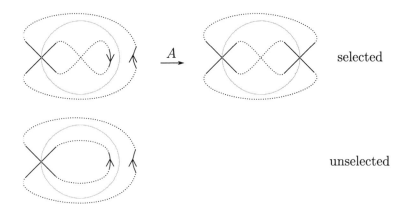

selected

unselected

FIGURE 9.35 Inverse A^{-1} of a half-twisted splice

Corollary 9.1 *Let P be a reduced knot projection. The following three conditions are equivalent.*
(1) $t(P) = 1$,
(2) $r(P) = 1$,
(3) $y(P) = 1$.

Proof 9.1 Suppose that (1) $t(P) = 1$. Then, there exists a simple closed curve intersecting at just two double points of P, as shown in Figure 9.27. Then, $y(P) \leq 1$ and $r(P) \leq 1$. Further, by the first statement of Proposition 9.1, $1 = t(P) \leq r(P), y(P)$. Thus, $t(P) = r(P) = y(P) = 1$, which implies (2) and (3). Then,

$$t(P) = 1 \Longrightarrow r(P) = 1 \text{ and } y(P) = 1.$$

Next, we suppose that (2) $r(P) = 1$. Because P is a reduced knot projection, $1 \leq t(P)$. Then, by the first statement of Proposition 9.1, $1 \leq t(P) \leq r(P) = 1$. Then,

$$r(P) = 1 \Longrightarrow t(P) = 1.$$

Suppose that (3) $y(P) = 1$. Because P is a reduced knot projection, $1 \leq t(P)$. Then, by the first statement of Proposition 9.1, $1 \leq t(P) \leq y(P) = 1$. Thus,

$$y(P) = 1 \Longrightarrow t(P) = 1.$$

Now we show an example for understanding of the difference among the four reductivities (Figs. 9.36 and 9.37). This example was introduced by Taniyama (cf. [2]). Let P be a knot projection, as shown in the leftmost figure in Figure 9.36. The double points of P are marked by circles or squares, which implies that there are exactly two types of double points by the symmetry of the knot projection. Thus, it is sufficient to consider only two kind of splices for the application of a splice to P. After applying the first splice (two cases)

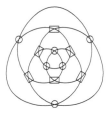 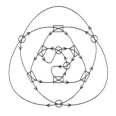 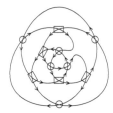

FIGURE 9.36 Knot projection (left) and two possibilities of A^{-1} (center, right)

to transform a knot projection into another, the center and rightmost figures shown in Figure 9.36 are obtained. These two figures are not reducible knot projections, which implies that $2 \le t(P) \le y(P), r(P)$. Further, by observing these two figures, we check every possibility for applying the next splice A^{-1}; we have $r(P) \ne 2$ and $y(P) \ne 2$. Then, $3 \le r(P), y(P)$. The reason for the condition $y(P) \ne 2$ can be obtained as follows. Suppose that $y(P) = 2$. Then, two splices are required to obtain a reducible knot projection, where the two splices are of type A^{-1}. One of the two splices A^{-1} is shown in Figure 9.36 (center or right). See the center or right of Figure 9.36. If the next splice is a Seifert splice, then we cannot have a single component curve, i.e., a knot projection. If the next splice is A^{-1}, then we cannot have a reducible knot projection. This is a contradiction. Thus, $y(P) \ne 2$.

On the other hand, it can be easily seen that $t(P) \le i(P) \le 2$. We can find two double points such that if two Seifert splices are applied to them, a reducible knot projection is obtained.

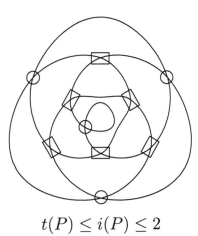

$$t(P) \le i(P) \le 2$$

FIGURE 9.37 $t(P) \le i(P) \le 2$

Here, we fix the values of $r(P)$ and $y(P)$. We can check whether a 3-gon of P is of type B. Then, $r(P) \leq 3$ and $y(P) \leq 3$ (cf. Section 9.4), which implies $r(P) = y(P) = 3$ using the above result. Alternatively, we can apply three splices of type A^{-1}, recursively, to obtain a reducible knot projection, as shown in Figure 9.38. The route is (top-left) $\to$ (top-right) $\to$ (bottom-right) $\to$ (bottom-left). Figure 9.38 also shows that $y(P) \leq 3$. The route is (top-left)

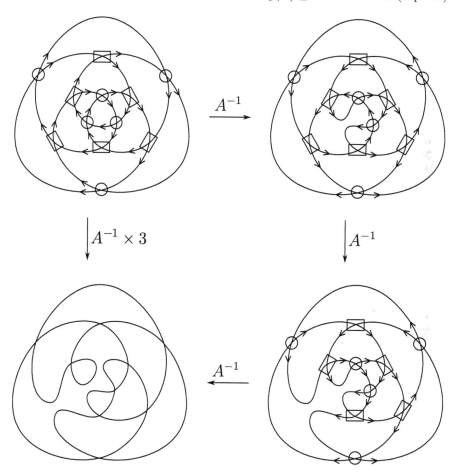

FIGURE 9.38 $r(P) \leq 3$ and $y(P) \leq 3$

to (bottom-left) in Figure 9.38.

To obtain $2 \leq t(P)$, we have already shown another method. In Chapter 4, for every knot projection Q, we define circle number $|\tau(Q)|$. For a knot projection Q, if $t(Q) = 1$, then there exists a simple closed curve intersecting at just two double points of P, as shown in Figure 9.27. Then, $2 \leq |\tau(Q)|$.

Therefore, for a reduced knot projection Q,

$$|\tau(Q)| = 1 \Longrightarrow 2 \leq t(Q).$$

Now we look back at the example of the leftmost figure of Figure 9.36, denoted by P. As shown in Figure 9.39, $|\tau(P)| = 1$, then $2 \leq t(P) \leq y(P), i(P), r(P)$.

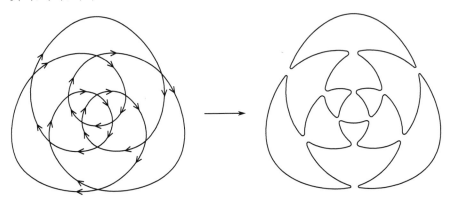

FIGURE 9.39 $|\tau(P)| = 1 \Longrightarrow 2 \leq t(P), y(P), r(P), i(P)$

9.6 KNOT PROJECTION WITH REDUCTIVITY TWO

Knot projections with $r(P) = 2$ are described below.

Theorem 9.5 ([2]) *Let P be a reduced knot projection such that there is no simple closed curve intersecting P at just two double points, as shown in Figure 9.27. There exists a simple closed curve intersecting P at just two or three double points, as shown in Figure 9.40 (1)–(5) if and only if $r(P) = 2$.*

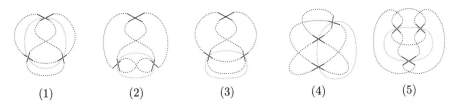

FIGURE 9.40 Knot projections with $r = 2$

Proof 9.2 Let P be a knot projection. Recall the necessary and sufficient condition, $r(P) = 1$. There exists a simple closed curve C intersecting P at just

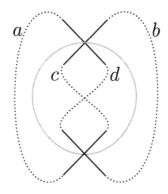

FIGURE 9.41 Dotted parts a, b, c, and d

two double points, as shown in Figure 9.41. Here, four dotted sub-curves are denoted by a, b, c, and d, as shown in Figure 9.41. Thus, as a step further, we consider one more half-twisted splice A, applied between two possible simple arcs. This operation A should eliminate the condition $r(P) = 1$ for a simple closed curve C. Then it is sufficient to show the following possibilities. Check Case 1–Case 3 to complete the proof.

Case 1. We apply A to two simple arcs that belong to a and b.

Case 2. We apply A to two simple arcs that belong to a and c. By symmetry, pairs (a, d), (b, c), (b, d) belong to this case. Note that there exist possibilities that mirror images appear, but every knot projection in Figure 9.40 is the same as its mirror image.

Case 3. We apply A to two simple arcs that belong to c and d.

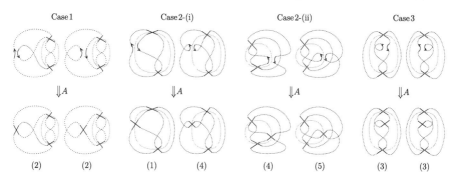

FIGURE 9.42 Cases 1–3

Knot projections with $i(P) = 2$, $y(P) = 2$, or $t(P) = 2$ can be defined [2] as follows. We introduce these results with no proof; for the proofs, see [2].

Theorem 9.6 ([2]) *Let P be a reduced knot projection such that there does not exist a simple closed curve intersecting P at just two double points of P, as shown in Figure 9.27. Then the following two conditions are equivalent:*
($\star$) $i(P) = 2$.
($\star\star$) There exists a simple closed curve intersecting P at just two or three double points, as shown in Figure 9.43.

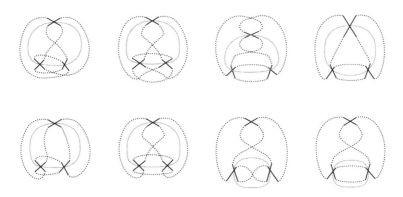

FIGURE 9.43 Reduced knot projection with $i(P) = 2$

Theorem 9.7 ([2]) *Let P be a reduced knot projection such that there does not exist a simple closed curve intersecting P at just two double points, as shown in Figure 9.44. Then the following two conditions are equivalent.*
($\star$) $y(P) = 2$.
($\star\star$) There exists a simple closed curve intersecting P at just two or three double points, as shown in Figure 9.44.

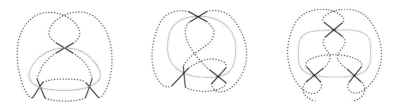

FIGURE 9.44 Knot projections with $y(P) \leq 2$

Theorem 9.8 ([2]) *Let P be a reduced knot projection such that there does not exist a simple closed curve intersecting P at just two double points, as shown in Figure 9.44. Then, the following two conditions are equivalent.*
($\star$) $t(P) = 2$.

(**) *There exists a simple closed curve intersecting at just two or three double points, as shown in Figure 9.43, Figure 9.44, and Figure 9.45.*

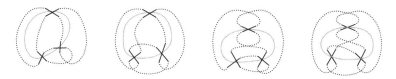

FIGURE 9.45 Knot projections with $t(P) \leq 2$

We do not obtain the proofs, but we shall show the key points. In order to prove Theorems 9.6, 9.7, and 9.8, we use two strategies. Refer to Figure 9.46. Let B be the inverse operation of a Seifert splice. Recall that a half-twisted splice is denoted by A.

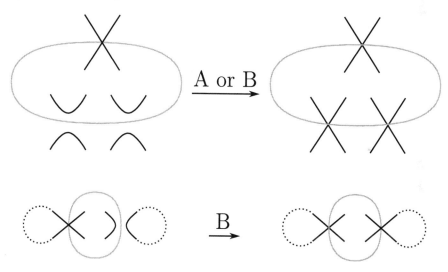

FIGURE 9.46 Half-twisted splice A and the inverse B of Seifert splice

1. (upper figure of Figure 9.46) By definition, $t(P)$, $y(P)$, and $i(P)$ are obtained by simultaneous splices. Then, if we apply two splices, each of which is A or B to a reducible knot projection, we have a knot projection P with its reductivity less than 3. Thus, we consider all possibilities for this case.

2. (bottom figure of Figure 9.46) Recall that there is no knot projection such that $i = 1$. Thus, we consider the last operation B for two component curves. We consider all the possibilities for this case.

9.7 TIPS

9.7.1 Tip I

Proposition 9.2 (well known) *Let $V(P)$ be the number of vertexes, $E(P)$ be the number of edges, and $F(P)$ be the number of faces of P. For every knot projection P,*

$$V(P) - E(P) + F(P) = 2.$$

Proof 9.3 • (Step 1). Remove a single face of P. The result of this operation is denoted by P'.

• (Step 2). Apply operations (I) or (II) to P' with $F \geq 2$ as follows.

(I) If there exists a vertex v such that v does not connect two edges i.e., a single edge e is connected to v, remove e and v, as shown in Figure 9.47.

FIGURE 9.47 Operation (I)

(II) Suppose that there exists no vertex satisfying the condition in case (I). In such a situation, choose an edge e such that e belongs to a boundary of a face f. Then, remove e and f as shown in Figure 9.48.

FIGURE 9.48 Operation (II)

If $F \geq 2$, we can apply either (I) or (II). As a result, we have $F = 1$, i.e., a single n-gon (n is a positive integer).

• (Step 3). For a single n-gon T, we have $V - E + F = 1$. Note that operations (I) and (II) do not change $V - E + F$. Together with (Step 1),

$$V(P) - E(P) + F(P) = 2.$$

9.7.2 Tip II

We can summarize the upper bounds of reductivities, as shown in the table (Figure 9.49). For example, if a knot projection has at least one 3-gon of type

C, we have $t(P) \leq 2$. This is because by the first statement of Proposition 9.1, $t(P) \leq i(P)$, and by the fourth statement of Proposition 9.1, $i(P) \leq 2$.

In this table, the second line for $r(P)$ is calculated by Shimizu [4]. The other values are calculated as shown in [2] except for $y(P) \leq 3$ for the existence of 3-gons of type B (Section 9.4). The symbol ∞ indicates that there exist no upper bounds, by Proposition 9.1, and there exists a knot projection P_m indicted by a nonnegative integer m such that $i(P_m) = 2m$. Blanks indicate the upper bounds are unknown for the corresponding case.

	weak 2-gon	strong 2-gon	A 3-gon	B 3-gon	C 3-gon	D 3-gon
r	1	2	2	3	3	4
t	1	2	2	3	2	4
y	1		2	3		
i	∞	2			2	

FIGURE 9.49 Numbers indicate known upper bounds. Blanks indicate unknown upper bounds.

9.8 OPEN PROBLEMS AND EXERCISES

1. (open [4]) For every knot projection P, $r(P) \leq 3$? Otherwise, find a counterexample: Can you find a knot projection with $r(P)=4$?

2. (open [2]) Can you find a counterexample for Question 9.3? Can you find a reduced knot projection with no 2-gons such that every 3-gon is of type D?

3. (open [2]) For every knot projection P, $t(P) \leq 2$? Otherwise, find a counterexample.

4. (open [2]) Can you find a knot projection with $t(P) = 3$?

5. Define a notion corresponding to multi-component spherical curves, i.e., link projections. Define operations corresponding to a splice from a link projection to another link projection. Then, check whether we need to mention to "simultaneously" or "recursively" in the versions of the definition of $t(P)$. That is, check whether the three definitions are equivalent or not.

   (a) $t'(P) = \min\{\text{number of splices to obtain a reducible knot projection}\}$,

(b) $t(P) = \min\{$number of splices applied simultaneously to obtain a reducible knot projection$\}$, and

(c) $t''(P) = \min\{$number of splices applied recursively to obtain a reducible knot projection$\}$.

6. Extend two notions of knot projection and their splices to link projections and their splices. Then, check whether the three definitions are equivalent.

(a) $i'(P) = \min\{$number of Seifert splices to obtain a reducible knot projection$\}$,

(b) $i(P) = \min\{$number of Seifert splices, applied simultaneously, to obtain a knot projection$\}$, and

(c) $i''(P) = \min\{$number of Seifert splices, applied recursively, to obtain a knot projection$\}$.

7. (open) Determine knot projections with reductivity $r(P) = 3$.

8. (open) Determine knot projections with reductivity $t(P) = 3$.

9. (open [2]) If a knot projection P has at least one strong 2-gon, can you find an integer $y(P) \le k$?

10. (open [2]) If a knot projection P has a 3-gon of type A, can you find k such that $i(P) \le k$?

11. (open [2]) If a knot projection P has a 3-gon of type B, can you find k such that $i(P) \le k$?

12. (open [2]) If a knot projection P has a 3-gon of type C, can you find k such that $y(P) \le k$?

13. (open [2]) If a knot projection P has a 3-gon of type D, can you find k such that $y(P) \le k$?

14. (open [2]) If a knot projection P has a 3-gon of type D, can you find k such that $i(P) \le k$?

Bibliography

[1] N. Ito and A. Shimizu, The half-twisted splice operation on knot projection, *J. Knot Theory Ramifications* **21** (2012), 1250112, 10pp.

[2] N. Ito and Y. Takimura, Knot projections with reductivity two, *Topology Appl.* 193 (2015), 290–301.

[3] N. Ito and Y. Takimura, Triple chords and strong (1, 2) homotopy, to appear in *J. Math. Soc. Japan.* **68** (2016), 637–651

[4] A. Shimizu, The reductivity of spherical curves, *Topology Appl.*, **196** (2015), part B, 860–867.

·

CHAPTER 10

Weak (1, 2, 3) homotopy

CONTENTS

10.1   Definitions 156
10.2   Strong (1, 2, 3) homotopy and the other triples 156
10.3   Definition of the first invariant under weak (1, 2, 3)
       homotopy 158
10.4   Properties of $W(P)$ 161
10.5   Tips 164
       10.5.1   The trivializing number is even 164
       10.5.2   One-half of the trivializing number is greater than
                or equal to the canonical genus of a knot
                projection 167
       10.5.3   If $tr(P) = c(P) - 1$, then P is a $(2, p)$-torus knot
                projection 172
10.6   Exercise 174

For refined Reidemeister moves, i.e., RI, strong RII, weak RII, strong RIII, and weak RIII, we have been considering equivalence relations, each of which is generated by two refined Reidemeister moves of a tuple of type (RI, y): $(RI, \emptyset)$, $(RI, strong\ RII)$, $(RI, weak\ RII)$, $(RI, strong\ RIII)$, and $(RI, weak\ RIII)$. For each equivalence relation, we have already determined the equivalence class that contains a trivial knot projection. We shall take one step further and consider the equivalence relations generated by three refined Reidemeister moves of type (RI, x, y): $(RI, strong\ RII, strong\ RIII)$, $(RI, strong\ RII, weak\ RIII)$, $(RI, weak\ RII, strong\ RIII)$, $(RI, weak\ RII, weak\ RIII)$. As shown by Ito–Takimura [3], all knot projections are related to a trivial knot projection under an equivalence relation (RI, x, y), as shown above, except for $(RI, weak\ RII, weak\ RIII)$. The equivalence relation corresponding to $(RI, weak\ RII, weak\ RIII)$ is called *weak (1, 2, 3) homotopy*. It is still not known which knot projection is equivalent to a trivial knot projection under weak (1, 2, 3) homotopy. Ito–Takimura [3] introduced the first invariant and showed that there exist infinitely many equivalence classes of knot projections under weak (1, 2, 3) homotopy.

10.1 DEFINITIONS

Let P and P' be knot projections.

1. If P and P' are related by a finite sequence generated by RI, weak RII, and weak RIII, P and P' are said to be weakly $(1, 2, 3)$ homotopic,

2. If P and P' are related by a finite sequence generated by RI, strong RII, and strong RIII, P and P' are said to be strongly $(1, 2, 3)$ homotopic,

3. If P and P' are related by a finite sequence generated by RI, weak RII, and strong RIII, P and P' are said to be $(1, w2, s3)$ homotopic,

4. If P and P' are related by a finite sequence generated by RI, strong RII, and weak RIII, P and P' are said to be $(1, s2, w3)$ homotopic.

10.2 STRONG (1, 2, 3) HOMOTOPY AND THE OTHER TRIPLES

Proposition 10.1 ([3]) *The following statements hold.*

(A) *For any two knot projections P and P', P and P' are strong $(1, 2, 3)$ homotopic.*

(B) *For any two knot projections P and P', P and P' are $(1, w2, s3)$ homotopic.*

(C) *For any two knot projections P and P', P and P' are $(1, s2, w3)$ homotopic.*

Proof 10.1 Recall that we have already shown the following fact (Reidemeister's theorem for knot projections, see Chapter 2).

Fact 10.1 *For any two knot projections P and P', P and P' are related by a finite sequence generated by RI, RII, and RIII.*

Thus, it is sufficient to show the following.

Lemma 10.1 ([3]) (1) *weak RIII is replaced with a finite sequence generated by strong RII and strong RIII.*

(2) *strong RIII is replaced with a finite sequence generated by strong RII and weak RIII.*

(3) *strong RII is replaced with a finite sequence generated by RI, weak RII, and strong RIII.*

(4) *weak RII is replaced with a finite sequence generated by RI, strong RII, and weak RIII.*

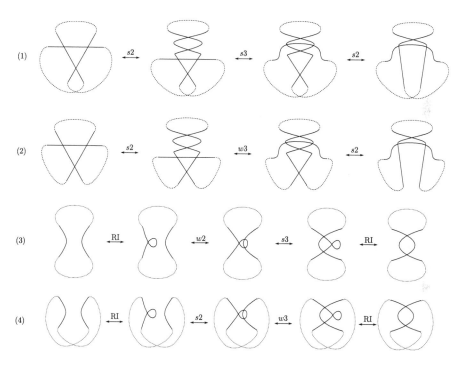

FIGURE 10.1 Symbol $s2$ (resp. $w2$) indicates strong RII (resp. weak RII) and symbol $s3$ (resp. $w3$) indicates strong RIII (resp. weak RIII)

(**Proof of Lemma 10.1.**) In Figure 10.1, the numbers (1)–(4) correspond to the number of the statements (1)–(4).
(**End of the proof of Lemma 10.1.**)

Now we show the statements.

Claim (A): By Lemma 10.1 (1), weak RⅢ is obtained by a finite sequence generated by strong RⅡ and strong RⅢ. Then, by Lemma 10.1 (4), weak RⅡ is generated by a finite sequence generated by RI, strong RⅡ, and strong RⅢ. Thus, using Fact 10.1, we have Claim (A).

Claim (B): By Lemma 3, strong RⅡ is obtained by a finite sequence generated by RI, weak RⅡ, and strong RⅢ. Then, by using Claim (A), we have Claim (B).

Claim (C): By Lemma 2, strong RⅢ is obtained by a finite sequence generated by strong RⅡ and weak RⅢ. Then, by using Claim (A), we have Claim (C).

10.3 DEFINITION OF THE FIRST INVARIANT UNDER WEAK (1, 2, 3) HOMOTOPY

The triple (RI, weak RⅡ, weak RⅢ) corresponding to weak (1, 2, 3) homotopy is quite different from the other triples containing RI. There exists a knot projection P such that P and a trivial knot projection are *not* equivalent under weak (1, 2, 3) homotopy. For example, a 7_4 projection shown in Figure 10.2 cannot be equivalent to a trivial knot projection under weak (1, 2, 3) homotopy. In order to prove the non-triviality of each 7_4 projection under weak (1,

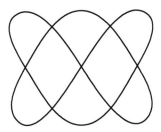

FIGURE 10.2 7_4 projection

2, 3) homotopy, we introduce the following invariant.

Definition 10.1 For a given knot projection P, we apply Seifert splice to every double point of P; then, we have an arrangement of a finite number of disjoint circles. The arrangement of circles on a sphere is called the *Seifert circle arrangement* and is denoted by $S(P)$. Every circle in $S(P)$ is called a *Seifert circle*. The number of circles is denoted by $s(P)$ and is called the *Seifert circle number* of P. Let $tr(P)$ be the trivializing number of P, and $c(P)$ be the number of double points. Here, the definition of the trivializing number is obtained from Chapter 6. The integer $W(P)$ is defined by

$$W(P) = tr(P) + s(P) - c(P) - 1.$$

Remark 10.1 One may feel that the representation of $W(P)$ should be simpler. This is because the number $(1+c(P)-s(P))/2$ is known as the canonical genus. Here, the canonical genus is denoted by $g(P)$. Thus, we also have

$$W(P) = tr(P) - 2g(P).$$

Theorem 10.1 ([3]) *Let P be a knot projection. Let $tr(P)$ be the trivializing number of P, $s(P)$ be the Seifert circle number of P, and $c(P)$ be the number of double points of P. Then,*

$$W(P) = tr(P) + s(P) - c(P) - 1$$

is invariant under weak (1, 2, 3) homotopy.

Proof 10.2 We have shown that $tr(P)$ is invariant under RI and weak RIII (cf. Chapter 6). It is easy to see that $c(P)$ is invariant under weak RIII. Moreover, it can be seen that $s(P)$ is invariant under weak RIII, as shown in Figure 10.3. Thus, it is sufficient to verify the invariance of $W(P)$ under RI and

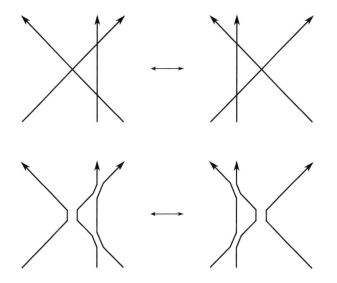

FIGURE 10.3 $s(P)$, which is invariant under weak RIII

weak RII. Let $1a$ be a single RI increasing the number of double points and $w2a$ be a single weak RII increasing the number of double points. The trivializing number $tr(P)$ changes by exactly $+2$ under $w2a$. This is because $w2a$ adds two chords that mutually intersect on a chord diagram of P, as shown in Figure 10.4. Thus, at least $tr(P)$ increases by at least 1. Here, using the fact that $tr(P)$ is always even, the increment is exactly $+2$.

Table 10.1 shows the invariance; hence, this completes the proof.

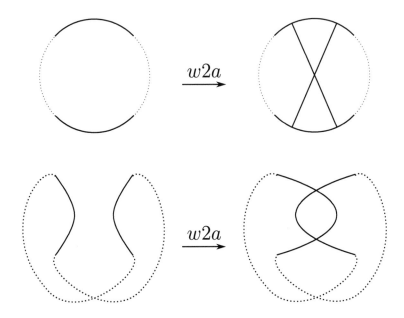

FIGURE 10.4 $w2a$ on a chord diagram

TABLE 10.1 Differences of $tr(P)$, $s(P)$, and $c(P)$

	$tr(P)$	$s(P)$	$c(P)$
$1a$	0	+1	+1
$w2a$	+2	0	+2

10.4 PROPERTIES OF $W(P)$

The invariant $W(P)$ has the following properties.

Theorem 10.2 ([3]) *Let P and P' be knot projections.*

1. *The integer $W(P)$ is even.*

2. *For a connected sum $P\sharp P'$, $W(P\sharp P') = W(P) + W(P')$ holds.*

3. *The inequality $0 \le W(P) \le c(P) - 1$ holds.*

4. *If $W(P) = c(P) - 1$, then P is a knot projection that appears as ∞.*

5. *For any non-negative even integer $2m$, there exists a non-trivial knot projection P such that $W(P) = 2m$. Further, we can select a prime knot projection P such that $W(P) = 2m$.*

Proof 10.3   1. From Remark 10.1, we can see that $s(P) - c(P) - 1$ is even. If another elementary proof is required to be obtained, we should consider the difference $s(P) - c(P) - 1$ under RI, RII, and RIII. From Proposition 10.1 (C), it is sufficient to verify that $s(P) - c(P)$ is 0 or changes by ± 2 under strong RII. From Figure 10.5, we can see that $s(P)$ is 0 or changes by ± 2 under strong RII, which implies that $s(P) - c(P) - 1$ is an even integer.

Thus, it is sufficient to show that $tr(P)$ is an even integer, which was shown by Hanaki [1, Theorem 1.6]. This proof is a bit lengthy and we will show it in Section 10.5.

2. By definition, $tr(P\sharp P') = tr(P) + tr(P')$ and $c(P\sharp P') = c(P) + c(P')$. It can be seen that $s(P\sharp P') = s(P) + s(P') - 1$. Then,

$$W(P\sharp P') = (tr(P) + tr(P')) + (s(P) + s(P') - 1) - (c(P) + c(P')) - 1$$
$$= (tr(P) + s(P) - c(P) - 1) + (tr(P) + s(P) - c(P) - 1)$$
$$= W(P) + W(P').$$

3. From Remark 10.1, we can see that $0 \le W(P)$ is equivalent to the statement $tr(P)/2 \ge g(P)$, which was shown by [2, Theorem 7.11]. The proof of this theorem is also a bit lengthy and, thus, we will prove it in Section 10.5. On the other hand, it can be easily seen that $W(P) = tr(P) + s(P) - c(P) - 1 \le c(P) + c(P) - c(P) - 1 = c(P) - 1$.

4. Suppose that $W(P) = c(P) - 1$. Then, from $tr(P) \le c(P) - 1$ and $s(P) \le c(P) + 1$, we can see that $tr(P) = c(P) - 1$ and $s(P) = c(P) + 1$. From [1, Theorem 1.11] (also see Section 10.5), we know that if $tr(P) = c(P) - 1$, then P is as shown in Figure 10.6. Thus, if P satisfies $tr(P) = c(P) - 1$ and $s(P) = c(P) + 1$, then P is a knot projection that appears, i.e., P is a knot projection having only one double point.

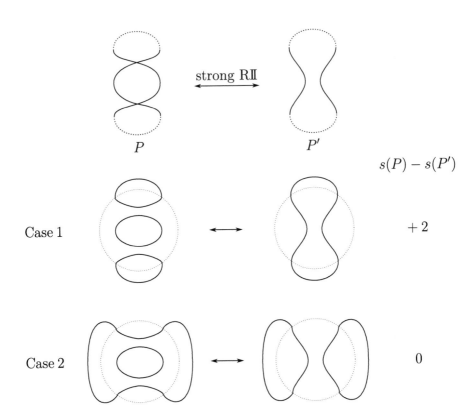

FIGURE 10.5 Difference $|s(P) - s(P')|$, which is 0 or 2 under a single strong R$\rm{I\!I}$ from P to P'

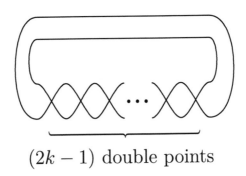

$(2k - 1)$ double points

FIGURE 10.6 Knot projection P with $tr(P) = c(P) - 1$

5. Note that $W(3_1) = 0$ and $W(7_4) = 2$ (Figure 10.2 and Figure 10.7). If we consider that a connected sum P consists of m copies of 7_4 and a single 3_1, then $W(P) = 2m$.

Next, we consider a prime knot projection $P(\alpha_1, \alpha_2)$ parameterized by two even positive integers α_1, α_2 ($\alpha_1 \leq \alpha_2$), as shown in Figure 10.7. Because the chord diagram is as shown in Figure 10.8, $tr(P(\alpha_1, \alpha_2)) =$

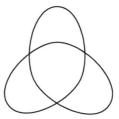

FIGURE 10.7 3_1 projection

α_1. We also have $s(P(\alpha_1, \alpha_2)) = \alpha_1 + \alpha_2 - 1$ and $c(P(\alpha_1, \alpha_2)) = \alpha_1 + \alpha_2$. Thus, $W(P(\alpha_1, \alpha_2)) = \alpha_1 - 2$.

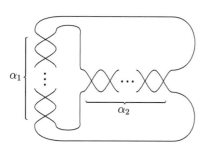
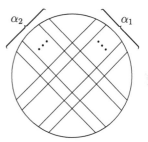

FIGURE 10.8 $P(\alpha_1, \alpha_2)$ with α_1, α_2 : even integers

Alternatively, we consider the following knot projections $P(\beta_1, \beta_2, \beta_3)$ parameterized by three odd positive integers $\beta_1, \beta_2, \beta_3$ ($3 \leq \beta_1 \leq \beta_2 \leq \beta_3$), as shown in Figure 10.9. $tr(P(\beta_1, \beta_2, \beta_3)) = \beta_1 + \beta_2$. $s(P(\beta_1, \beta_2, \beta_3)) = \beta_1 + \beta_2 + \beta_3 - 1$ and $c(P(\beta_1, \beta_2, \beta_3)) = \beta_1 + \beta_2 + \beta_3$. Thus, $W(P(\beta_1, \beta_2, \beta_3)) = \beta_1 + \beta_2 - 2$, $W(3_1) = 0$, and $W(7_4) = 2$, which show the claim.

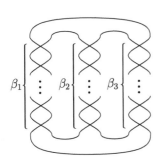
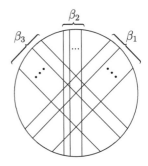

FIGURE 10.9 $P(\beta_1, \beta_2, \beta_3)$ with odd integers β_1, β_2, and β_3

10.5   TIPS

10.5.1   The trivializing number is even

In this section, we prove Hanaki's Theorem:

Theorem 10.3 ([1]) *Any trivializing number is an even integer.*

Before starting the proof, we obtain the definition of a *teardrop disk* for a knot projection (cf. Chapter 3).

Definition 10.2 For a knot projection P, we start at a double point d, proceed along P, and return to d. If the sub-curve obtained from this trajectory is a simple closed curve, i.e., its inner region is a disk D, then we call D a *teardrop disk*, as shown in Figure 10.10. The double point d is called a *teardrop-origin*.

Proof of Theorem 10.3. Let P be a knot projection. Let $c(P)$ be the number of double points of P. Let CD_P be a *chord diagram*, as defined in Chapter 6. A *sub-chord diagram* is also as defined in Chapter 6. From a geometrical observation, if some chords are not considered in CD_P, the result is called a *sub-chord diagram* of CD_P. By definition, there exists a sub-chord diagram CD' consisting of $c(P) - tr(P)$ chords in CD_P such that $\bigotimes(CD') = 0$ (for the definition of $\bigotimes(CD')$, see Chapter 6, Definition 6.1). The chord diagram CD' is denoted by $CD_{tr(P)}$. A chord diagram CD with $\bigotimes(CD) = 0$ is called a *trivial chord diagram*. Recall the definition of $tr(P)$ is

$$tr(P) = \min\{\text{number of chords eliminated until a trivial chord diagram}$$
$$\text{as a sub-chord diagram of } CD_P \text{ is obtained}\}.$$

By definition, $CD_{tr(P)}$ is a sub-chord diagram of CD_P with maximal number of chords among the sub-chord diagrams, each of which satisfies $\bigotimes(CD) = 0$.

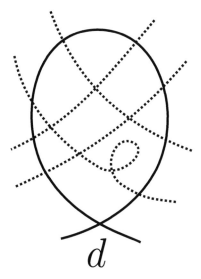

FIGURE 10.10 Teardrop disk

If we identify $CD_{tr(P)}$ with a disk, each chord in $CD_{tr(P)}$ divides the disk into two disks. For $CD_{tr(P)}$, there exists a chord that divides it into two disks such that one of the two disks contain no chords. Such a chord is called an *outermost chord*. Note that every outermost chord corresponds to a teardrop disk. Let us take an outermost chord in $CD_{tr(P)}$. If the corresponding teardrop disk contains a smaller teardrop disk, we retake the smaller teardrop disk. Following the same procedure, we can choose an outermost chord c_1 corresponding to a teardrop disk δ such that δ does not contain any other teardrop disk. Then, if the sub-curves $l_1(\delta), l_2(\delta), \ldots, l_m(\delta)$ pass through δ, each $l_i(\delta)$ has no self-intersections in δ, as shown in Figure 10.11. This is because if there exists l_i such that it is an $(1, 1)$-tangle having a double point (for the definition of *tangles*, see Chapter 9, Definition 9.2), then the $(1, 1)$-tangle has at least one teardrop disk, as shown in Figure 10.12. Here, note that every knot projection has at least two teardrop disks because there exist at least two outermost chords in a chord diagram.

Recall that the outermost chord c_1 is retained after the $c(P) - tr(P)$ chords are eliminated. We consider the two chords corresponding to the two double points that are $\partial\delta \cap l_i(\delta)$. By definition, the two chords and c_1 should intersect. These two chords should be eliminated when we eliminate $tr(P)$ chords. That is, for every $l_i(\delta)$, we should eliminate a pair of chords corresponding to $\partial\delta \cap l_i(\delta)$. In the end, we eliminate the outermost chord c_1. This process, that is, eliminating even chords (double points) and c_1 (the teardrop-origin), can be considered as applying a finite sequence generated by $\overline{\Delta}$, RI, RII, and RIII (cf. *pulling operation* in Chapter 2).

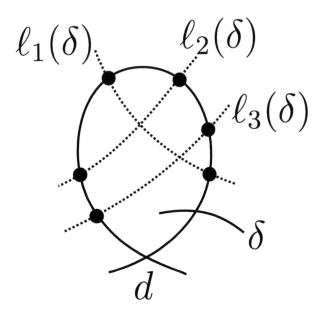

FIGURE 10.11 Teardrop disk δ and $l_i(\delta)$

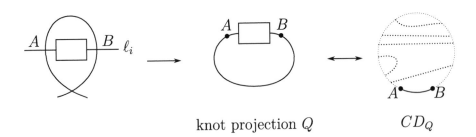

FIGURE 10.12 Sub-curve l_i, knot projection Q, and its chord diagram CQ_Q

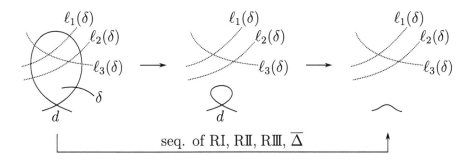

FIGURE 10.13 Resolutions of double points between the boundary of δ and $l_i(\delta)$ interpreted by applying a finite sequence generated by RI, RII, and RIII

Next, we consider an outermost chord c_2 after the elimination of odd numbers of chords corresponding to c_1, double points between $l_1(\delta), l_2(\delta), \ldots, l_m(\delta)$, and the boundary δ by a finite sequence generated by $\overline{\Delta}$, RI, RII, and RIII. For c_2, we repeat the above process. Finally, all $tr(P)$ chords and chords corresponding to teardrop-origins are eliminated, i.e., $c(P)$ chords are eliminated. Note that for each step, the number of chords contributing to $tr(P)$ is even. Thus, $tr(P)$ is even. □

10.5.2 One-half of the trivializing number is greater than or equal to the canonical genus of a knot projection

Before proving the claim, we provide relevant definitions

Definition 10.3 Let us consider m knot projections that may intersect with each other and every knot projection has an orientation. This is traditionally called an m-component (oriented) *link projection D*. For every double point, the local replacement (the replacement of a sufficient small disk) is defined as shown in Figure 10.14. If we apply a Seifert splice to every double point, a finite number of circles are obtained. The number of circles is denoted by $s(D)$.

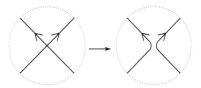

FIGURE 10.14 Seifert splice

The proof follows that provided in [2, Theorem 7.11]. Recall that if a chord diagram does not contain ⊗, we call it a trivial chord diagram. By the definition of the trivializing number $tr(P)$, $c(P) - tr(P)$ corresponds to the cardinality of a maximal set of chords of a trivial chord diagram. A chord diagram consists of a circle and chords (corresponding to the paired points). Here, a circle and its inner region can be regarded as a disk. Note that the corresponding $c(P) - tr(P)$ chords divide the disk into $c(P) - tr(P) + 1$ disks. Thus, if we apply Seifert splices to $c(P) - tr(P)$ double points (corresponding to the $c(P) - tr(P)$ chords), $c(P) - tr(P) + 1$ knot projections D are obtained that may intersect with each other. For D, recall that the number of Seifert circles is denoted by $s(D)$ after the application of a Seifert splice to every double point of D. Because $s(P) = s(D)$, it is sufficient to show that $s(D) \geq c(P) - tr(P) + 1$, which follows from Lemma 10.2.

Lemma 10.2 *For an n-component link projection Q, $s(Q) \geq n$.*

In [2], the authors use an important fact to knot theory, i.e., Yamada's Theorem, as follows: "for a link, the minimum number of Seifert circles is the braid index" [4]. However, they mentioned that "This may be shown by induction on n." Let us consider the proof by an induction on n. Before starting the proof for an n-component link, we show the special case when $n = 1, 2$.

Lemma 10.3 (1) *A knot projection has at least one Seifert circle.*
(2) *A two-component link projection has at least two Seifert circles.*

Proof of Lemma 10.3.
(1) If we apply a Seifert splice to every double point of a knot projection, we obtain a finite number of circles with no double points; at least one circle is obtained after the application of the Seifert splice. This is because a knot projection is not empty; thus, at least one arc should belong to a Seifert circle (e.g., in Figure 10.15, the chosen arc is marked by a base point).
(2) Let P be a given two-component link projection. We apply a Seifert splice to every self-intersection for each component of P. Then, we have a link projection with at least two components where each component has no self-intersection; this is denoted by $P(n)$ with n double points. Now we consider the following operations (O1) and (O2).
(O1): Suppose that $n > 0$. Because each component is a simple closed curve, we can find a pair of two double points, as shown in Figure 10.16 (left). There can be two cases for the pair.

1. Case 1: An orientation of a simple closed curve that coincides with that of another.

2. Case 2: An orientation of a simple closed curve that does not coincide with that of another.

We apply a Seifert splice to each double point belonging to a pair. For each case, after the application of the Seifert splices, we obtain a two-component link projection with $(n - 2)$ double points and denote it by $Q(n - 2)$ (Figure 10.16, right).

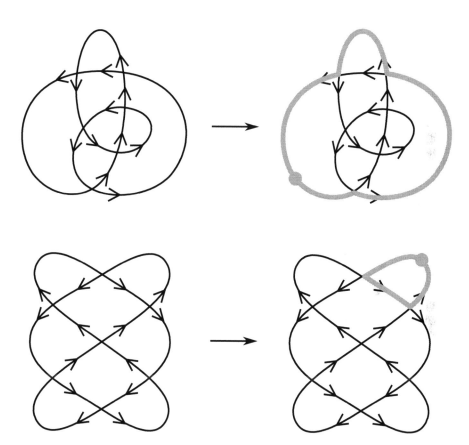

FIGURE 10.15 Knot projection having at least one Seifert circle

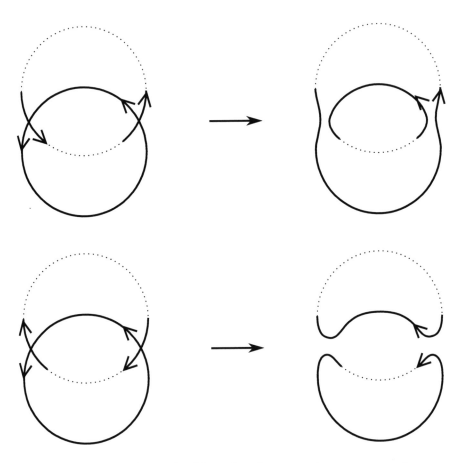

FIGURE 10.16 Pair of two double points between two components

(O2): Suppose that after the completion of operation (O1), a link projection $Q(n - 2)$ with $(n - 2)$ double points is obtained, where the number of components of $P(n)$ is equal to that of $Q(n - 2)$. We apply a Seifert splice to every self-intersection of each component of $Q(n - 2)$ and we obtain a link projection $P(m)$ with m double points and at least two components. Note that $m \le n - 2 < n$.

We repeat operations (O1) and (O2) alternatively. Because the number of double points is finite, the number of the operations are applied is also finite; finally, we will have a link projection $P(0)$ with no double points and at least two components. By the definition of (O1) and (O2), the number of circles in $P(0)$ is greater than or equal to 2. By definition, $P(0)$ is Seifert circle arrangement $S(P)$ and the number of circles in $S(P)$ is $s(P)$.

Thus, the claim is verified. □

Example 10.1 Two examples for a process of applying operations (O1) and (O2) are shown in Figs. 10.17 and 10.18.

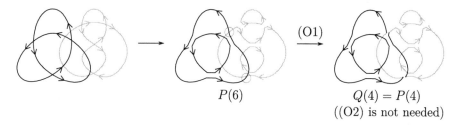

$P(6)$ (O1) $Q(4) = P(4)$
((O2) is not needed)

FIGURE 10.17   Example 1: Process (O1) and (O2)

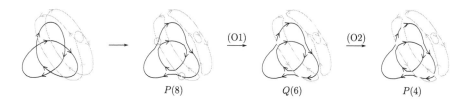

$P(8)$ (O1) $Q(6)$ (O2) $P(4)$

FIGURE 10.18   Example 2: Process (O1) and (O2)

Proof of Lemma 10.2. Consider induction on n, which denotes the number of components. First, for the case when $n = 1$, Lemma 10.3 (1) implies that the claim is true. Second, we assume that the claim is true for n is true and consider the case $n + 1$.

Let D denote an $n + 1$-component link diagram. For D, we choose n-components arbitrarily, which are denoted by $D(n)$. If we eliminate $D(n)$ in the link diagram, a single component remains, which is denoted by l_{n+1}.

Now we apply a Seifert splice to each double point of $D(n)$ and obtain the Seifert circle arrangement $S(D(n))$. By the induction assumption, there exist at least n simple closed curves in $S(D(n))$.

Moreover, we apply a Seifert splice to each double point of l_{n+1}, when other components are eliminated and obtain the Seifert circle arrangement $S(l_{n+1})$. Suppose that the number of simple closed curves in $S(l_{n+1})$ is m. We label the simple closed curves in $S(l_{n+1})$ and denote them by $T_1, T_2, \ldots, T_m$.

Now we consider the intersections between $S(D(n))$ and T_i, where these intersections are obtained from the original link diagram D. All simple closed curves in $S(D(n))$ are denoted by $U_1, U_2, \ldots, U_k$. We choose a single component U_j from $S(D(n))$. T_i and U_j consist a two-component link projection with N_{ij} intersections between T_j and U_j, which is denoted by $P(N_{ij})$ and where $N_{ij} > 0$. Recall the operations (O1) and (O2) on a two-component link projection where each component has no self-intersection. We apply (O1) and (O2) to the double points of $P(N_{ij})$ and apply (O1) and (O2) iteratively in a manner similar to that shown in the proof of Lemma 10.3 until we obtain $P(0)$. By the definition of (O1) and (O2), the number of components does not decrease.

Recall that $i = 1, 2, \ldots, m$ and $j = 1, 2, \ldots, k$. After we obtain $P(0)$ from $P(N_{ij})$, if necessary, we renumber the elements of the set of simple closed curves in $P(0)$, $\{T_{i'} \mid (1 \le i' \le m)\} \setminus \{T_i\}$, and $\{U_{j'} \mid (1 \le j' \le k)\} \setminus \{U_j\}$. Then, we choose a two-component link projection having at least one double point where each component has no self-intersection and apply (O1) and (O2) iteratively to the link projection. Following the same procedure, we finally obtain a link projection with no double points, which is denoted by $D(0)$.

Note that the application of operations (O1) and (O2) does not decrease the number of components; thus, $D(0)$ has at least $(n + 1)$ components $(*)$. Here, if we do not ignore the over-/under information of the double points of D, D is denoted by $P(D)$. By definition, $D(0)$ has a Seifert circle arrangement $S(P(D))$. $(*)$ implies that $S(P(D))$ has at least $(n + 1)$ Seifert circles.

This completes the induction, which implies the proof. □

10.5.3 If $tr(P) = c(P) - 1$, then P is a $(2, p)$-torus knot projection

Suppose that $tr(P) = c(P) - 1$. Then, $c(P) - tr(P) = 1$. From Section 10.5.1, we know that if there exist two chords that do not mutually intersect each other, then $c(P) - tr(P) \ge 2$. Hence, every chord intersects with the other chords. The remaining possibilities are as shown in Figure 10.19. Here, the chord diagram of type shown in Figure 10.19 has odd chords because every chord of a chord diagram of a knot projection is intersects even chords (this can be easily verified; e.g., see Figure 10.20).

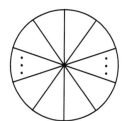

FIGURE 10.19 Chord diagram where each chord intersects the other chords

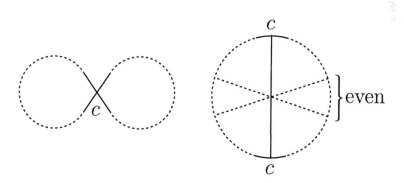

FIGURE 10.20 Double point c and the corresponding chord

10.6 EXERCISE

1. Verify that every chord of a chord diagram of a knot projection intersects $2m$ chords $(m \in \mathbb{Z}_{\geq 0})$.

Bibliography

[1] R. Hanaki, Pseudo diagrams of knots, links and spatial graphs, Osaka J. Math. **47** (2010), 863–883.

[2] A. Henrich, N. Macnaughton, S. Narayan, O. Pechenik, and J. Townsend, Classical and virtual pseudodiagram theory and new bounds on unknotting numbers and genus, *J. Knot Theory Ramifications*, **20** (2011), 625–650.

[3] N. Ito and Y. Takimura, Strong and weak (1, 2, 3) homotopies on knot projections, *Internat. J. Math.* **26** (2015), 1550069, 8pp.

[4] S. Yamada, The minimal number of Seifert circles equals the braid index of a link, *Invent. Math.* **89** (1987), 347–356.

CHAPTER 11

Viro's quantization of Arnold invariant

CONTENTS

11.1 Reconstruction of Arnold invariant J^- 177
11.2 Generalization 185
11.3 Quantization 188
11.4 Knot projection on a sphere 190
11.5 Exercises 195

Quantization of Arnold invariant was archived by Viro [1].

11.1   RECONSTRUCTION OF ARNOLD INVARIANT J^-

In this section, we review Viro's reconstruction of Arnold invariant J^-. Let C be an oriented plane curve that is the image of a generic immersion from an oriented circle into a plane. Because the curve C is the image of a generic immersion, $C = f_C(S^1)$, where for each $t \in S^1$, $f_C(t) = x \in C \subset \mathbb{R}^2$. Now we select and fix a point $a \in \mathbb{R}^2 \setminus C$. Then, we consider the map $C \to S^1$ such that

$$g_a : x \mapsto \frac{x - a}{|x - a|}.$$

Then, we have the map $g_a \circ f_C : S^1 \to S^1$.

We consider the number of times we go around S^1. Thus, if we go around S^1 once, then $g_a \circ f_C(S^1)$ will go around S^1 m times for $m \in \mathbb{Z}$ (cf. Exercise 1). It is elementary to show that integer m is determined by the pair (a, C). We denote integer m by $\text{ind}_C(a)$.

Note that $\mathbb{R}^2 \setminus C$ or $S^2 \setminus P$ consists of a finite number of disjoint regions. In this chapter, regardless of the situation, considering a knot projection C on a plane or a knot projection P on a sphere, any region is also called a 2-*simplex* (cf. Definition 5.4). Then, a 2-simplex is denoted by σ.

Now, we go back to $\text{ind}_C(a)$. By definition, it is easy to see that $\text{ind}_C(a)$ does not depend on the choice of a point in σ. Thus, for every $a \in \sigma$, $\text{ind}_C(a)$

is denoted by $\mathrm{ind}_C(\sigma)$. Then, for a given C, we consider every 2-simplex σ and define the map $\{\sigma\}_{\sigma \subset \mathbb{R}^2 \setminus C} \to \mathbb{Z}$ such that

$$\sigma \mapsto \mathrm{ind}_C(\sigma).$$

For a given C, obtaining $\{\mathrm{ind}_C(\sigma)\}_{\sigma \subset \mathbb{R}^2 \setminus C}$ is called *Alexander numbering* and an example is shown in Figure 11.1. The calculation of $\mathrm{ind}_C(\sigma)$ is simple by

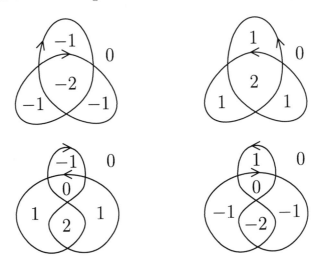

FIGURE 11.1 Examples of Alexander numbering

considering the Seifert splices of the double points of C as follows.

For a given oriented plane curve C, we apply a Seifert splice to every double point. After the applications of Seifert splices to C, the arrangement of oriented circles on the plane is called the *Seifert circle arrangement*, denoted by $S(C)$. Every circle in $S(C)$ is called a *Seifert circle* (cf. Chapter 3). Then we have

$$\mathbb{R}^2 \setminus S(C) = \coprod_{R:\text{connected component}} R \qquad \text{(a disjoint union)}.$$

Each R is called a *region* of the Seifert circle arrangement. We assign an integer for each region $S(C)$, as follows. First, we assign 0 to the exterior region that contains ∞ of $\mathbb{R}^2$. Second, for any $a \in \mathbb{R}^2 \setminus S(C)$, let $s_1, s_2, \ldots, s_l$ be the Seifert circles surrounding a. If a Seifert circle s_i surrounds a, then we assign ± 1 to this circle. If a Seifert circle s_i is oriented counterclockwise (resp. clockwise), then $+1$ (resp. -1) is assigned to the circle as $\epsilon(s_i) = +1$ (resp. -1). Then, the number $\sum_{i=1}^{l} \epsilon_i$ is denoted by $\mathrm{ind}'_C(a)$. It is easy to see the following:

Proposition 11.1 *Let C be an oriented plane curve. Let σ be a 2-simplex, R*

be a region of $S(C)$, and two functions ind *and* ind' *be as defined above. For* $a \in \mathbb{R}^2 \setminus (C \cup S(C))$, *except for the small neighborhoods of the double points,*

$$\mathrm{ind}_C(a) = \mathrm{ind}'_C(a).$$

Proof of Proposition 11.1. Let p_∞ be a point in $\sigma \subset \mathbb{R} \setminus C$ containing ∞. Let a be a point in $\mathbb{R}^2 \setminus C$. Consider an oriented segment s from a to p_∞. Without loss of generality, we can suppose that point p_∞ satisfies the condition that there is no tangency between s and C (suppose that, if necessary, we can retake a). If a fragment of s passes through C from left to right, we add $+1$. If not, we add -1. By using the definition of $\mathrm{ind}_C(a)$, the sum of all the signs is equal to $\mathrm{ind}_C(a)$ such that $\mathrm{ind}_C(a) = \mathrm{ind}_C(\sigma)$ for $a \in \sigma$, where σ is a 2-simplex. Thus, for any $\sigma \subset \mathbb{R}^2 \setminus C$, $\mathrm{ind}_C(\sigma)$ is assigned according to the following rule. If we consider the orientation of any oriented edge of C to be from south to north and the 2-simplex on the east is considered to have index $i = \mathrm{ind}_C(a)$, then the 2-simplex on the west will have index $i + 1$, as shown in Figure 11.2.

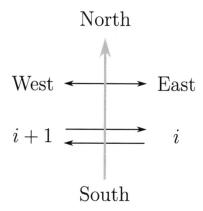

FIGURE 11.2   Two neighboring simplexes with indices

Then, for every double point, we assign indices as shown in Figure 11.3. It is easy to see that the arrangement of indices in the left figure of Figure 11.4 is the same as that in the right figure for every point $a \in \mathbb{R}^2 \setminus (C \cup S(C))$ except for the small neighborhoods of the double points. The arrangement of the indices appearing in the left figure of Figure 11.4 corresponds to $\mathrm{ind}'_C(a)$. □

Example 11.1 Figure 11.5 shows the examples of Seifert circle arrangements with indices corresponding to Figure 11.1. This is also obtained by Alexander numbering.

By using Proposition 11.1, we have the following.

Proposition 11.2 ([1]) *Let C be a plane curve. The configuration of integers* $\mathrm{ind}_C(a)$ *($a \in \mathbb{R}^2 \setminus C$) is invariant under* weak RII *and weak RIII.*

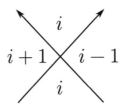

FIGURE 11.3 Arrangement of indices in the neighborhood of a double point of C, where i is an index $\mathrm{ind}_C(a)$.

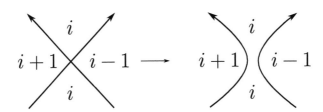

FIGURE 11.4 Arrangement of $\mathrm{ind}_C(a)$ (left) and of $\mathrm{ind}'_C(a)$ (right)

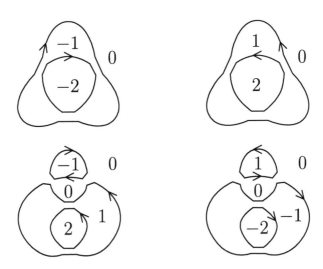

FIGURE 11.5 Examples: Seifert splices and Alexander numbering

Proof of Proposition 11.2. By Proposition 11.1, we have already obtained $\text{ind}_C(a) = \text{ind}'_C(a)$. Thus, we check the difference between the Seifert circle arrangements before and after the application of weak RII or weak RIII; it is easy to check the invariance, as shown in Figure 11.6. □

By Proposition 11.1, for the rest of this section, we freely use the identification $\text{ind}_C(a)$ with $\text{ind}'_C(a)$. We can see that $\text{ind}'_C(a)$ does not depend on the choice of a. Thus, for any region R of the Seifert circle arrangement $S(C)$, we shall denote $\text{ind}'_C(a)$ ($a \in R$) by $\text{ind}_{S(C)}(R)$. For any region R of $S(C)$, let $\chi(R)$ be the Euler characteristic of R.

Theorem 11.1 ([1]) *Let C be a plane curve and let R be a region of a Seifert circle arrangement $S(C)$.*

$$J^-(C) = 1 - \sum_{R \subset \mathbb{R}^2 \setminus S(C)} (\text{ind}_{S(C)}(R))^2 \chi(R).$$

Proof of Theorem 11.1. Recall the definition of J^- (cf. Chapter 7) as follows. J^- is a function on the set of the unoriented plane curves in a regular homotopy class. The value of J^- is an odd integer.

Let K_i be a sequence of plane curves as shown in Figure 11.7. Set

$$J^-(K_0) = -1, \ J^-(K_i) = -3(i-1).$$

For a given plane curve C, the rotation number C is $\text{rot}(C)$. $J^-(C)$ can be defined recursively from K_i such that it satisfies the following condition:

• (condition) J^- changes by -2 (resp. $+2$) under $s2a$ (resp. $s2b$) and does not change under weak RII and RIII, where $s2a$ and $s2b$ are defined by Figure 11.8.

Now, we check the formula mentioned in the statement. First, we check the changes in

$$\sum_{R \subset \mathbb{R}^2 \setminus S(C)} (\text{ind}_{S(C)}(R))^2 \chi(R)$$

under $s2a$ and $s2b$. Considering the orientation of C, two cases are required to be checked as shown in Figure 11.9.

Consider the difference (the value after the application of $w2a$) − (the value before the application of $w2a$) for each case, Case 1 and Case 2.

• Case 1 (upper line of Figure 11.9): Refer to Figure 11.10.

(a) $(i+2)^2 - 2(i+1)^2 + i^2 = 2$.

(b) $(i+2)^2 - 2(i+1)^2 + i^2 = 2$.

• Case 2 (lower line of Figure 11.9): Refer to Figure 11.11.

(a) $(i+2)^2 - 2(i+1)^2 + i^2 = 2$.

(b) $(i+2)^2 - 2(i+1)^2 + i^2 = 2$.

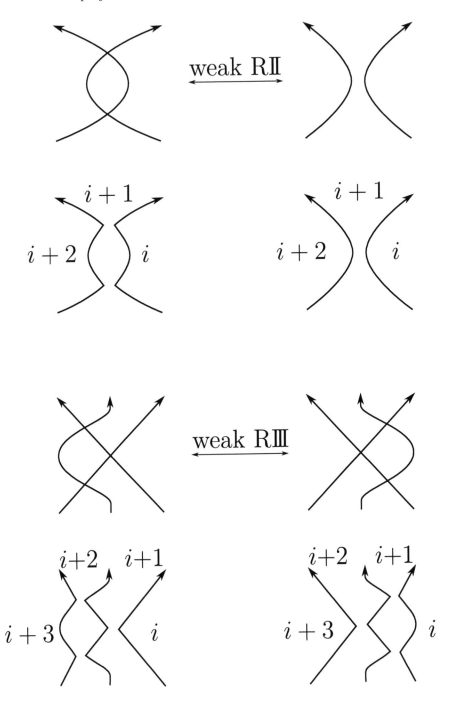

FIGURE 11.6 Seifert circle arrangements with indices under weak RII and weak RIII

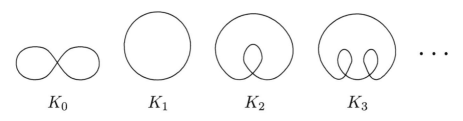

FIGURE 11.7 $K_i\ (i \in \mathbb{Z}_{\geq 0})$

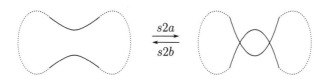

FIGURE 11.8 $s2a$ and $s2b$

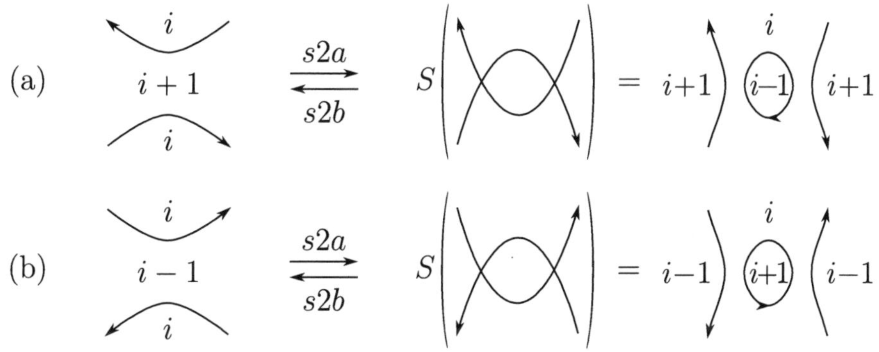

(a)

(b)

FIGURE 11.9 Arrangement of indices before and after the applications of $s2a$ or $s2b$

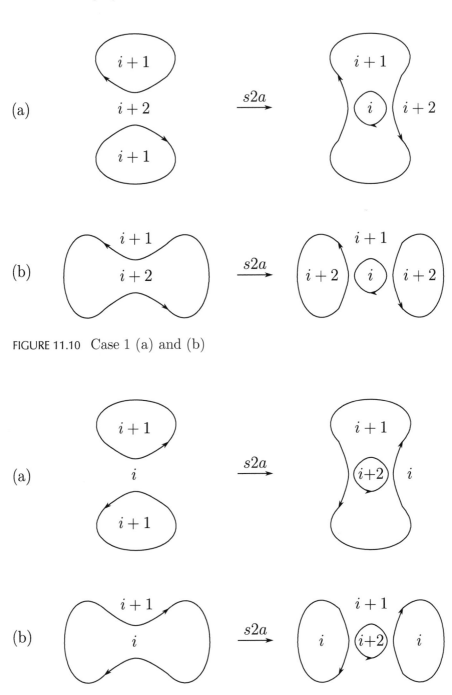

FIGURE 11.10 Case 1 (a) and (b)

FIGURE 11.11 Case 2 (a) and (b)

As a result, the number $\sum_{R\subset\mathbb{R}^2\setminus S(C)}(\mathrm{ind}_{S(C)}(R))^2\chi(R)$ increases by 2 under *s2a*.

Next, we check its invariance under weak RII, strong RIII, and weak RIII. By Proposition 11.2, $S(C)$ does not change under weak RII and weak RIII and thus, it is sufficient to check its invariance under strong RIII. Consider (a)–(c) shown in Figure 11.12.

(a) $(i+3)^2 - 3(i+2)^2 + 3(i+1)^2 - i^2 = 0$.

(b) $(i+3)^2 - 3(i+2)^2 + 3(i+1)^2 - i^2 = 0$.

(c) $(i+3)^2 - 3(i+2)^2 + 3(i+1)^2 - i^2 = 0$.

Then, we check the value of K_i ($i \in \mathbb{Z}_{\geq 0}$). Refer to Figure 11.13. The value of

$$\sum_{R\subset\mathbb{R}^2\setminus S(K_i)}(\mathrm{ind}_{S(K_i)}(R))^2\chi(R) :$$

1. $i = 0$ (resp. $i = 1$), the value is 2 (resp. 1).

2. $i \geq 2$, the value is $1^2(-(i-2)) + 2^2(i-1) = 3i - 2$.

Then,

$$1 - \sum_{R\subset\mathbb{R}^2\setminus S(K_i)}(\mathrm{ind}_{S(K_i)}(R)^2)\chi(R)$$

is -1 ($i = 0$) and $-3(i-1)$ ($i \geq 1$). □

11.2 GENERALIZATION

Following the previous sections, we quote Viro's generalization of Arnold invariant J^-.

Recall that the Euler characteristic satisfies the following. For any two sets $A, B \subset \mathbb{R}^2$,

$$\chi(A \cup B) = \chi(A) + \chi(B).$$

Similarly, for any plane curve C and for any two regions R_1, R_2 of $S(C)$,

$$\chi(R_1 \cup R_2) = \chi(R_1) + \chi(R_2) - \chi(R_1 \cap R_2).$$

Let R be a region of $S(C)$. For any R, we consider a constant function on R with value $\mathrm{ind}_{S(C)}(R)$. Recall that $S(C) = \coprod_{R:\text{connected component}} R$. The function is naturally extended to a locally constant function on $\mathbb{R}^2 \setminus S(C)$. This locally constant function is denoted by $\mathrm{ind}_{S(C)}(x)$ on $\mathbb{R}^2 \setminus S(C)$. Then, we set the notation of an integral (by Viro) such that

$$\int_{\mathbb{R}^2\setminus S(C)} \mathrm{ind}_{S(C)}(x)d\chi(x) = \sum_{R\subset\mathbb{R}^2\setminus S(C)} \mathrm{ind}_{S(C)}(R)\chi(R).$$

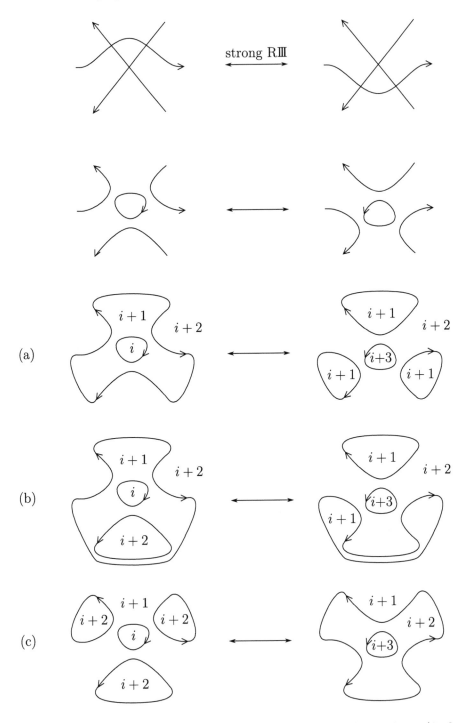

FIGURE 11.12 Strong RⅢ (1st line), corresponding Seifert splices (2nd line), and cases indices (a)–(c)

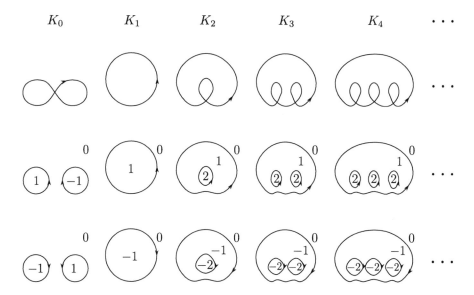

FIGURE 11.13 Indices for K_i

By using this notation, we represent the sum $\sum_{R \subset \mathbb{R}^2 \setminus S(C)} \operatorname{ind}_{S(C)}(R)\chi(R)$ by

$$\int_{\mathbb{R}^2 \setminus S(C)} \operatorname{ind}_{S(C)}(x)d\chi(x). \tag{11.1}$$

Then, it is easy to extend (11.1) to the following, by replacing a locally constant function $\operatorname{ind}_{S(C)}(x)$ with $(\operatorname{ind}_{S(C)}(x))^r$:

$$\int_{\mathbb{R}^2 \setminus S(C)} (\operatorname{ind}_{S(C)}(x))^r d\chi(x) = \sum_{R \subset \mathbb{R}^2 \setminus S(C)} (\operatorname{ind}_{S(C)}(R))^r \chi(R).$$

This formula is denoted by $M_r(C)$. For example: for case $r = 1$, we have the following:

Proposition 11.3 ([1]) *Let C be a plane curve and let $\operatorname{rot}(C)$ be the rotation number of C. Then,*

$$\operatorname{rot}(C) = M_1(C).$$

Proof of Proposition 11.3. Recall Theorem 11.1. We know that the difference of $M_2(C)$ under strong RII or strong RIII is

$$(i + 2)^2 - 2(i + 1)^2 + i^2 \tag{11.2}$$

$$(i + 3)^2 - 3(i + 2)^2 + 3(i + 1)^2 - i^2, \tag{11.3}$$

respectively.

Then, we can easily see that the difference of $M_r(C)$ under strong RII or strong RIII is

$$(i+2)^r - 2(i+1)^r + i^r \tag{11.4}$$

$$(i+3)^r - 3(i+2)^r + 3(i+1)^r - i^r, \tag{11.5}$$

respectively.

For example, setting $r = 1$ implies the invariance under RII and RIII.

Finally, we check the values of K_i with an orientation. The plane curve K_i with a nonnegative rotation number i is denoted by K_i^+ and with a negative rotation number i is denoted by K_i^-.

1. $i = 0$: $M_1(K_0^+)$ is 0.

2. $i = -1$: $M_1(K_{-1}^-)$ is -1.

3. $i = 1$: $M_1(K_1^+)$ is 1.

4. $i \geq 2$: $M_1(K_i^+)$ is $1 \cdot (-(i-2)) + 2(i-1) = i$.

5. $i \leq -2$: $M_1(K_i^-)$ is $(-1) \cdot (-(|i|-2)) + (-2) \cdot (|i|-1) = -|i|$.

To summarize, $M_1(\cdot)$ is i for K_i^+ and $-|i|$ for K_i^-. This completes the proof. □ By definition, we have the following:

Theorem 11.2 ([1]) *For each positive integer r, M_r is invariant under weak RII and weak RIII.*

Theorem 11.2 obtains an infinite family of invariants of plane curves under weak RII and weak RIII.

11.3 QUANTIZATION

The integer $M_r(C)$ of a plane curve C $(r = 1, 2, \ldots)$ results in a "quantization" that is similar to knot quantum invariants. It has been observed that each quantum knot polynomial that is invariant with its value in $\mathbb{Z}[q, q^{-1}]$ has a property as follows: If set $q = e^{ih}$, each coefficient of h^r is a finite-type invariant of links.

Viro defines a polynomial as follows. For a plane curve C, a formal power series $P_C(h)$ is considered such that

$$P_C(h) = \sum_{r=0}^{\infty} \frac{M_r(C)h^r}{r!}.$$

Viro obtains the following formula:

$$P_C(h) = \int_{\mathbb{R}^2 \setminus S(C)} e^{\mathrm{ind}_{S(C)}(x)h} d\chi(x). \tag{11.6}$$

(Proof of the equality of 11.6.)

$$\int_{\mathbb{R}^2\setminus S(C)} e^{\operatorname{ind}_{S(C)}(x)h} d\chi(x) = \int_{\mathbb{R}^2\setminus S(C)} \sum_{r=0}^{\infty} \frac{(\operatorname{ind}_{S(C)}(x))^r h^r}{r!} d\chi(x)$$

$$= \sum_{R\subset\mathbb{R}^2\setminus S(C)} \sum_{r=0}^{\infty} \frac{h^r (\operatorname{ind}_{S(C)}(R))^r \chi(R)}{r!}$$

$$= \sum_{r=0}^{\infty} \frac{h^r}{r!} \sum_{R\subset\mathbb{R}^2\setminus S(C)} (\operatorname{ind}_{S(C)}(R))^r \chi(R)$$

$$= \sum_{r=0}^{\infty} \frac{h^r}{r!} \int_{\mathbb{R}^2\setminus S(C)} (\operatorname{ind}_{S(C)}(x))^r d\chi(x)$$

$$= \sum_{r=0}^{\infty} \frac{h^r}{r!} M_r(C)$$

$$= P_C(h).$$

□

Now, consider a small perturbation such as RⅡ or RⅢ between two plane curves; then, there exists a curve C with a single self-tangency point or triple point. Then, the curve C is called a *singular plane curve*.

In the following, we say that an inverse self-tangency point p on a singular plane curve C has degree i, if the minimal number i satisfies the following condition: there exists a small perturbation C' of C and a point p' such that $\operatorname{ind}_{C'}(p') = i$, which can be moved to be arbitrarily close to p by retaking C'. Similar to the inverse self-tangency points, in the following, we say that a

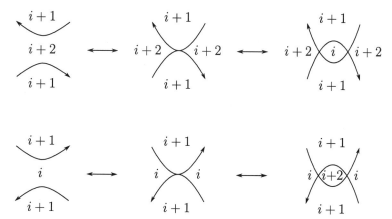

FIGURE 11.14   Inverse self-tangency point with degree i

triple point p on a singular plane curve C has degree i if the minimal number

i satisfies the following condition: there exists a small perturbation C' of C and a point p' such that $\text{ind}_{C'}(p') = i$, which can be moved to be arbitrarily close to p by retaking C'.

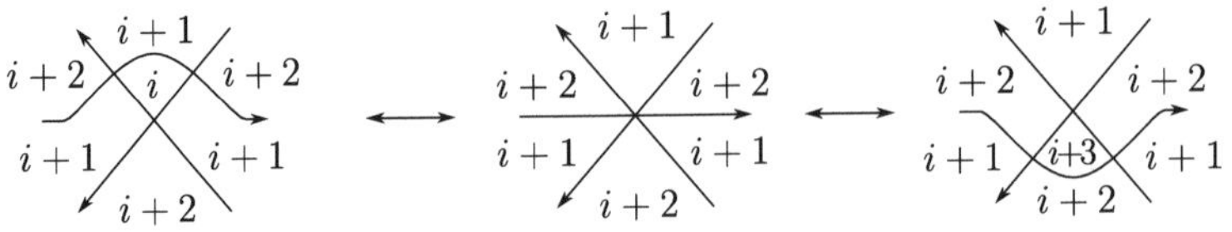

FIGURE 11.15 Triple point with index i

Recall that each $M_r(C)$ changes by $(i+2)^r - 2(i+1)^r + i^r$ (resp. $(i+3)^r - 3(i+2)^r + 3(i+1)^r - i^r$) under strong RII (resp. RIII) at an inverse self-tangency point (resp. a triple point) with degree i. Thus, $P_C(h)$ changes by $e^{ih}(e^h - 1)^2$ under inverse self-tangency at an inverse self-tangency point with degree i whereas it changes by $e^{ih}(e^h - 1)^3$ under a triple point crossing at triple point with degree i.

Setting $q = e^h$, $P_C(h)$ can be changed to $P_C(q)$ as follows:

$$P_C(q) = \int_{\mathbb{R}^2 \setminus S(C)} q^{\text{ind}_{S(C)}(x)} d\chi(x).$$

$P_C(q)$ changes by $q^i(q-1)^2$ under strong RII at an inverse self-tangency point with degree i and changes by $q^i(q-1)^3$ under RIII at a triple point with degree i.

11.4 KNOT PROJECTION ON A SPHERE

One may think that it is possible to obtain the counterpart of $M_r(C)$ on S^2. Following Viro's paper [1], we define $M_r(P)$ in this section. Let P be a knot projection on S^2. Recall that there exists a natural embedding from $\iota : \mathbb{R}^2 \to S^2$ (cf. Chapter 2). Let C be a plane curve depending on $x \in S^2 \setminus P$ such that $\iota(C) = P$ if x is considered to be the infinity point ∞. For x, let $w_P(x)$ be the rotation number of a plane curve C. For a given oriented knot projection P, we apply a Seifert splice to every double point. After the applications of the Seifert splices to P, the arrangement of the oriented circles on the sphere is called the *Seifert circle arrangement* (cf. Chapter 10). Note that $S^2 \setminus S(P) = \coprod_{R:\text{connected component}} R$. Each R is called a *region* of the Seifert circle arrangement $S(P)$. By definition, it is easy to see that $w_P(x)$ does not depend on the choice of point x in R. Thus, for every $x \in R$, $w_P(x)$ is denoted by $w_P(R)$. Recall the behavior of $\text{ind}_C(x)$. Compare the values $\text{ind}_C(x)$ and $w_P(x)$, as shown in Figure 11.16. Because we have already calculated the differences of $\text{ind}_{S(C)}(x)$ under strong RII, weak RII, strong RIII, and weak RIII, it is easy to obtain the differences of $-w_P(x)/2$ (Exercise 2). Here, we consider

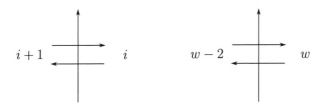

FIGURE 11.16 Difference of $\mathrm{ind}_C(x)$ (left) and difference of $w_P(x)$ (right)

the J^- invariant for knot projections on S^2. There is a well-known fact as shown in the following.

Fact 11.1 (Arnold) *Let P be a knot projection on S^2. Let C_P be a plane curve obtained by arbitrarily choosing a point in $S^2 \setminus P$ as an infinite point ∞ of $\mathbb{R}^2$. Let $\mathrm{rot}(C_P)$ be the rotation number of C_P. Then,*

$$J^-(C_P) + \frac{\mathrm{rot}(C_P)^2}{2}$$

does not depend on the choice of the infinite point.

Following Viro's theory [1], to prove Fact 11.1, we state Lemmas 11.1 and 11.2.

Lemma 11.1 ([1]) *Let ι be an embedding $\mathbb{R}^2 \to S^2$ and we identify x with $\iota(x)$. Let $\iota(C) = P$. For any $x \in \mathbb{R}^2 \setminus C$,*

$$\mathrm{ind}_C(x) = -\frac{1}{2}(w_P(x) - \mathrm{rot}(C)).$$

Proof of Lemma 11.1. We have already shown that the local behaviors of $\mathrm{ind}_C(x)$ and $-\frac{1}{2}(w_P(x) - \mathrm{rot}(C))$ coincide, as shown in Figure 11.16. Thus, it is sufficient to compare the values of ind_C and w_P between the unbounded region $R_0 \subset \mathbb{R}^2 \setminus C$ and $\iota(R_0)$. Then, we notice that the difference is $\mathrm{rot}(C)$ whereas $\mathrm{ind}_C(x) = 0$ ($x \in R_0$) (e.g., see Figure 11.17). □

From Lemma 11.1, using the identification $\mathrm{ind}_C(x)$ with $\mathrm{ind}'_C(x)$ (cf. Proposition 11.1), for any $x \in \mathbb{R}^2 \setminus S(C)$, we define $w'_P(\iota(x))$ by

$$w'_P(\iota(x)) = -2\,\mathrm{ind}'_C(x) + \mathrm{rot}(C).$$

By retaking the infinite point, we extend $w'_P(\iota(x))$ to the function on $S^2 \setminus S(P)$ and the extended function is denoted by $w'_P(y)$ ($y \in S^2 \setminus S(P)$). In the following, we freely use the identification $w_P(x)$ with $w'_P(x)$. An embedding ι such that $\iota(\mathbb{R}^2) = S^2$ induces a one-to-one correspondence between each

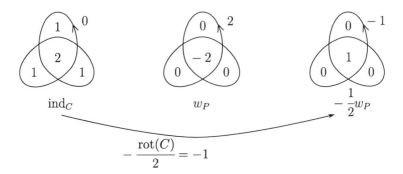

FIGURE 11.17 Example: differences between ind_C and w_P

connected component $R' \subset \mathbb{R}^2 \setminus S(C)$ and each connected component R of $S^2 \setminus S(P)$. Then, let

$$w_P(R) = -2\,\mathrm{ind}_C(R) + \mathrm{rot}(C)$$

and let

$$\int_{S^2 \setminus S(P)} (w_P(x))^r \, d\chi(x) = \sum_{R \subset S^2 \setminus S(P)} w_P(R)^r \chi(R).$$

Lemma 11.2 ([1])

$$\int_{S^2 \setminus S(P)} w_P(x) d\chi(x) = 0.$$

Proof of Lemma 11.2. Recall that the local behavior of $-\frac{1}{2}w_P(x)$ is equivalent to that of $\mathrm{ind}_C(x)$ (see Figure 11.16). Thus, it is sufficient to check the differences under strong RII and strong RIII because the above equality $-\frac{1}{2}(w_P(x) - \mathrm{rot}(C)) = \mathrm{ind}_C(x)$ (Lemma 11.1) implies invariances under weak RII and weak RIII.

Now we set $i = -\frac{1}{2}w_P(x)$. By the proof of Proposition 11.2, for strong RII, the difference is

$$(i+2) - 2(i+1) + i = 0$$

and for strong RIII, the difference is

$$(i+3) - 3(i+2) + 3(i+1) - i = 0.$$

Thus, $\frac{1}{2}\int_{S^2 \setminus S(P)} w_P(x) d\chi(x)$ is invariant under RII and RIII.

Here, let T_1 be a trivial knot projection on S^2 (a simple closed curve on S^2) and T_2 be a knot projection that appears as ∞ on S^2. It is easy to see that $\int_{S^2 \setminus S(T_i)} w_P(x) d\chi(x) = 0$ ($i = 1, 2$), which implies the proof. □

Remark 11.1 Recall that for a plane curve C, the definition of $M_r(C)$ is as follows. Here, let $S(C)$ be the Seifert circle arrangement of C and R be a region of $S(C)$.

$$M_r(C) = \int_{\mathbb{R}^2 \backslash S(C)} (\text{ind}_{S(C)}(x))^r d\chi(x) = \sum_{R \subset \mathbb{R}^2 \backslash S(C)} (\text{ind}(x))^r \chi(R).$$

Let R be a region of $S(P)$. Then, similarly, let

$$M_r(P) = \int_{S^2 \backslash S(P)} \left(-\frac{w_P(x)}{2} \right)^r d\chi(x).$$

The above formulae are defined for the case when r is a positive integer. Now we extend the definition for the case when $r = 0$. For a constant c,

$$\int_{S^2 \backslash S(P)} c\chi(x)$$

is defined by

$$\sum_{R \subset S^2 \backslash S(P)} c\chi(R) \quad (= 2 \cdot c).$$

This is a natural extension. Thus, we define

$$\int_{\mathbb{R}^2 \backslash S(C)} c\chi(x)$$

by

$$\sum_{R \subset \mathbb{R}^2 \backslash S(C)} c\chi(R) \quad (= c \cdot 1).$$

Let ι be an embedding from $\mathbb{R}^2$ to S^2. Let C be a plane curve and P be a knot projection on S^2 such that $\iota(C) = P$. Let $w_C(x)$ be the function $w_C : \mathbb{R}^2 \backslash C \to \mathbb{Z}$ such that $w_C(\iota(x)) = w_P(y)$ for $\iota(x) = y$. For the unbounded region $R_0 \subset \mathbb{R}^2 \backslash S(C)$, there exists a point $p \in S^2$ and a region $R_\infty \subset S^2 \backslash S(P)$ such that $\iota(R_0)$ is identified with $R_\infty \backslash \{p\}$. Then, $w_C(R_0) = w_P(R_\infty)$ and for $R \ (\neq R_0)$, $w_C(R) = w_P(\iota(R))$. Let

$$\int_{\mathbb{R}^2 \backslash S(C)} (w_C(x))^r = \sum_{R \subset \mathbb{R}^2 \backslash S(C)} (w_C(R))^r \chi(R).$$

Lemma 11.3 ([1]) *Under the above convention,*

$$\int_{\mathbb{R}^2 \backslash S(C)} (w_C(x))^r d\chi(x) = \int_{S^2 \backslash S(P)} (w_P(x))^r d\chi(x) - (\text{rot}(C))^r.$$

Proof of Fact 11.1. First, from Lemma 11.1, we have

$$\int_{\mathbb{R}^2\setminus S(C)} (\mathrm{ind}_{S(C)}(x))^2 d\chi(x) = \int_{\mathbb{R}^2\setminus S(C)} \left(\frac{1}{2}\mathrm{rot}(C) - \frac{1}{2}w_C(x)\right)^2 d\chi(x)$$

$$= \left(\frac{1}{4}\mathrm{rot}(C)\right)^2 - \frac{1}{2}\mathrm{rot}(C)\int_{\mathbb{R}^2\setminus S(C)} w_C(x)d\chi(x)$$

$$+ \frac{1}{4}\int_{\mathbb{R}^2\setminus S(C)} (w_C(x))^2 d\chi(x).$$

Here, we use

$$\int_{\mathbb{R}^2\setminus S(C)} d\chi(x) = 1$$

for the first term.

Next, from Lemma 11.3 and Lemma 11.2, we obtain the second term on the right-hand side of the above formula

$$\frac{1}{2}\mathrm{rot}(C)\int_{\mathbb{R}^2\setminus S(C)} w_C(x)d\chi(x) = \frac{1}{2}\mathrm{rot}(C)\int_{S^2\setminus S(P)} w_P(x)d\chi(x) - \frac{1}{2}(\mathrm{rot}(C))^2$$

$$= -\frac{1}{2}(\mathrm{rot}(C))^2.$$

Therefore,

$$\int_{\mathbb{R}^2\setminus S(C)} (\mathrm{ind}_{S(C)}(x))^2 d\chi(x) = \frac{1}{4}(\mathrm{rot}(C))^2 + \frac{1}{2}(\mathrm{rot}(C))^2 - \frac{1}{4}(\mathrm{rot}(C))^2$$

$$+ \int_{S^2\setminus S(P)} (w_P)^2 d\chi(x).$$

Further,

$$J^-(C) + \frac{1}{2}(\mathrm{rot}(C)) = 1 - \int_{\mathbb{R}^2\setminus S(C)} (\mathrm{ind}_C(x))^2 d\chi(x) + \frac{1}{2}(\mathrm{rot}(C))^2$$

$$= 1 - \frac{1}{4}\int_{S^2\setminus S(P)} (w_P(x))^2 d\chi(x).$$

This completes the proof. □

The formula

$$J^-(C_P) + \frac{(\mathrm{rot}(C_P))^2}{2} = 1 - \frac{1}{4}\int_{S^2\setminus S(P)} (w_P(x))^2 d\chi(x)$$

is denoted by $J_S^-(C_P)$, which implies an invariant of knot projections on S^2 under weak RII and RIII.

11.5   EXERCISES

1. Describe $g_a \circ f_C$ by the notion of the degree map. For example, if you know the notion of homology, you can consider the map $H(S^1) \to H(S^1)$.

2. Obtain immediate definitions of functions on S^2 corresponding to $\mathrm{ind}'_C(x)$ and $\mathrm{ind}_{S(C)}(x)$.

Bibliography

[1] O. Viro, Generic immersions of the circle to surfaces and the complex topology of real algebraic curves. *Topology of real algebraic varieties and related topics,* 231–252, Amer. Math. Soc. Transl. Ser. 2, 173, *Amer. Math. Soc., Providence, RI,* 1996.

Index

$(2, 2m + 1)$-torus knot projection, 139
2-simplex, 177
3-colorable, 58
RⅢ-chord, 81
strong RⅢ, 55
weak RⅢ, 55
a-move, 39, 102
b-move, 31, 39, 102
n-gon, 30
x-disk, 108

additivity, 39, 80
Alexander numbering, 178
ambient isotopy, 4
arc, 30, 66, 109
arc for knot diagram, 59
arrow, 85
averaged invariant, 103
avoidable set, 126

base point, 11, 85
based arrow diagram, 84, 85
box, 97
Box Rule, 97

charge, 127, 129
chord, 74
chord diagram, 73, 164
chord diagram of a knot projection, 74
circle arrangement, 40
circle number, 29, 40
connected sum, 66, 67, 82
crossing, 57

discharging, 127, 129, 130

edge, 1, 2, 126

endpoints of a tangle, 125
equivalence class, 54, 55
equivalence relation, 54

face, 126

Gauss word, 73

half-twisted splice, 122
head, 84

innermost chord, 82
interval, 125
invariant, 57, 75
isomorphism (Gauss word), 73
isomorphism (oriented Gauss word), 84

knot, 1, 4
knot diagram, xi, 1, 2
knot diagram on a plane, 3
knot diagram on a sphere, 3
knot projection, xiii, 4, 101

letter, 73
link projection, 167
local replacement, 122

negative crossing, 85
negative move, 102
non-Seifert splice, 44
nugatory double point, 44, 122, 124

oriented Gauss word, 84
oriented letter, 84
Östlund's question, xiii
outermost chord, 165

Perko pair, xi
piecewise linear knot, 4

plane curve, 17, 22, 101, 103
plane isotopy, 7
pointed knot diagram, 85
positive crossing, 57, 85
positive knot, 58
positive knot diagram, 58
positive move, 102
positive resolution, 57
prime knot projection, 67
pulling operation, 165

reduced knot projection, 46
reducible disk, 47
reducible knot projection, 44, 124
refined Reidemeister move, 55, 155
region (knot projection), 66
region (Seifert circle arrangement),
          178, 190
regular homotopy, 7
Reidemeister move, 55
Reidemeister's Theorem, 8
relation, 54
rotation number, 17

Seifert circle, 158, 178
Seifert circle arrangement, 158, 178,
          190
Seifert circle number, 158
Seifert splice, 23, 44, 121
simple arc, 97, 109
simple teardrop disk, 21
singular plane curve, 189
sphere isotopy, 7
splice, 44, 122
stereographic projection, 3
strong 2-gon, 37, 102
strong 3-gon, 79, 101, 102
sub-arrow diagram, 85
sub-chord diagram, 74, 164
swelling, 109

tail, 84
tame knot, 4
tangle, 125, 165
tangle on a sphere, 125
teardrop disk, 20, 21, 164

teardrop-origin, 20, 164
trivial chord diagram, 86, 164
trivial knot projection, 8

unavoidable set, 125, 126
underlying sub-curve, 79

vertex, 1, 2

weak 2-gon, 37, 102
weak 3-gon, 101, 102
weak (1, 2, 3) homotopy, 155
weak (1, 3) homotopy, 57
wild knot, 4

Printed and bound by CPI Group (UK) Ltd, Croydon, CR0 4YY

24/10/2024

01778284-0003